Mathematics As Known
To
The Vedic Saṁhitās

Sri Garib Das Oriental Series No. : 169

Mathematics As Known To The Vedic Saṁhitās

M.D. PANDIT

Sri Satguru Publications
A Division of
Indian Books Centre
Delhi—India

Published by:
SRI SATGURU PUBLICATIONS
Indological & Oriental Publishers,
A Division of
INDIAN BOOKS CENTRE
40/5, Shakti Nagar,
Delhi-110007
INDIA

First Edition : Delhi 1993

ISBN 81-7030-368-0

Typeset by:
ABC *for* DTP
6, Rajdhani Enclave, Delhi-34. Ph. 7183087

Printed at:
D.K. Fine Art Press,
Ashok Vihar. Delhi-110052

PRINTED IN INDIA

|| **Vedapuruṣāya namaḥ** ||

In total surrender
To
Vedapuruṣa

with an earnest prayer to open the doors of
His highest knowledge of the secrets of the universe
for the emancipation of humanity from
the bondage of ignorance

Foreword

We have great pleasure in associating ourselves with the publication of *Mathematics As Known To The Vedic Saṁhitās by* Dr. M.D. Pandit. The work is the culmination of author's concentrated efforts to investigate into Vedic Mathematics. A book on Vedic Mathematics is usually associated with fast mental calculations by the use of the *Sūtras. Sūtras* are not everything as Dr. Pandit points out. The author makes an earnest attempt to understand the scientific mind of the Vedic people. Mathematics is the Queen of all Sciences. An understanding of the mathematics of the Vedic people will, therefore, shed a light on their scientific temper. By restricting himself to the Saṁhitās the author has made some interesting and useful contributions to the study of mathematics. By a skilful employment of symbols of present day mathematics with Vedic words, the author has attempted to explain the Vedic Mathematics. Interestingly, the author points out how the Vedic people could have had two intrepretations of Zero - one as a number and the other as a "place value substituted for absence of rank". The idea of counting numbers in two different ways—ascending and descending orders as found in the Saṁhitās—has been well brought out. "The word nava-*daśa* (for 19) implies the starting point of count in *daśa* (10) and ascending is done by nine (*nava*) steps to arrive at 19. *Ekonaviṁśati* or *ekānnaviṁśati* indicates counting from *viṁśati* (20) and going down by one step to 19". Taking advantage of the linguistic intrepretation of numbers, the author has also introduced the notion of simple and compound numbers. For example, 1 to 9 are simple while 10, 11, 12 etc., are compound numbers. This is perhaps typical of Sanskrit language. It seems there is still much scope for the study of Vedic Mathematics. Dr. Pandit has shown the way. With the recent awareness of the use of Sanskrit in general and Vedic Mathematics in particular in computer science, the present work is very valuable.

In a work like this one has to rely on the Vedic texts only, as Dr. Pandit himself points out; there is no way of testing the accuracy of the literary evidence provided by the Vedas.

Dr. M.D. Pandit is an extremely devoted scholar and is dedicated to the work on Vedic Mathematics. We have had many valuable discussions on the subject. His enthusiasm and self-confidence have overwhelmed us. He has restricted himself only to the arithmatical aspect of Vedic Mathematics. We hope, the other aspects, such as algebraic, geometric etc. will be considered by him in his later works.

Dr. V.N. Jha
Centre of Advanced Study in Sanskrit,
University of Poona

Dr. T.T. Raghunathan
Department of Mathematics
University of Poona

Preface

I

We encounter with mathematical numbers at the very outset when we start studying the Sanskrit language. Every student of Sanskrit language has to imbibe well first its systax. Of all the four types of syntaxes found in the different languages of the world, viz. agglutinative, inflexional with the two varieties of internal and external inflection, positional or isolating and the polysynthetic or incorporating types of syntaxes, the Sanskrit language exhibits the inflexional—and there too the external inflection predominantly—type of syntax. The Sanskrit language thus, according to Schlege[1], displays "an organic" character, "the words of which were built round modifiable roots by means of intimately linked inflection". The result of this highly rigid and complicated nature of the inflection of the Sanskrit words has been that besides the linguistic meaning they convey, they also convey the correct gender, number and mutual relations so accurately that there is practically no chance of any confusion in understanding the correct meanings of the words. Thus in Sanskrit, the inflexional category viz. what is called as the *pratyaya* in Pāṇini's grammar is a compulsory category. This is the most important principle, underlying the Pāṇinian analysis of the Sanskrit language, as Patañjali in his *bhāṣya* on the Pāṇinian *sūtra,* 1.2.45. (*arthavad adhātur apratyayaḥ prātipadikam*) puts it; *cf.* Patañjali on 1.2.45: *pratyayena nityasaṁbandhāt. nitya—sambandhāv etāv arthau prakṛtiḥ pratyaya iti. pratyayena nityasambandhāt kevalasya prayogo na bhaviṣyati.*

It can be seen, therefore, that of all the six grammatical categories called *vyavasitas* by Patañjali, viz. *dhātu, pratyaya, āgama, ādeśa, prātipadika* and *nipāta*[2], the category of *pratyaya* plays the most important role in building up the syntax of the language and thereby in conveying the correct meaning of the words. As Vaiyākaraṇabhūṣanasāra puts it: *prakṛtipratyayau saha artham brūtaḥ; tayoḥ pratyayārthasya prādhānyam.*

Because of the pivotal role that *pratyaya* plays in signifying the meaning, it is invested with different powers. It thus signifies the *liṅga* or gender, the *vacana* or the number, *vibhakti* or the mutual *kāraka*-relations between words, the *kāla* or the tense in the case of *dhātus* and lastly, the *svara* or the accent in the Vedic Sanskrit. In other words, all these meanings of gender, number, relations, tense and accent reside or dwell in the *pratyayas*[3], in the language and consequently in Pāṇini's grammar. It is because of the different powers of signifying the meaning, which are invested in the category of *pratyaya,* that the use of single word like *devấu* signifies the meanings of 'masculine, dual, nom or acc., and accent' simultaneously; it is because of this power of the *pratyaya* that the word *gácchati* signifies simultaneously the meanings of 'singular, 3rd person, present tense and its initial position in the beginning of the foot of a verse'. The inflectional nature of the Sanskrit language has made the syntax so rigid with a potential capacity of signifying the meaning as accurately as possible that there is no chance of misunderstanding a sentence, so far as the above-mentioned meanings are concerned.

To put the whole discussion symbolically. If F represents the form used in the language, N represents the nucleus or base or *prakṛti* and S stands for the suffix or *pratyaya* applied to the *prakṛti,* the most general formula for a finished word usable in the language will be:

$$F = N + S$$

But this is a representation of the pure mechanical procedure or analysis. It is devoid of any semantic implication. To make the representation fully faithful to its use in the language, we have to put life in it. The life can be put by investing the S with the powers of signifying the gender, number, *kāraka*-relations, accent, tense etc. And, we represent the above formula in a more refined way as follows:

$F = N + S^{g+n+r+a+m}$ for a nominal form, in which g = gender, n = number, r = kāraka - relations, a = accent and m for meaning. For a verbal form, the formula will be:

$F = N + S^{t+n+r+a+m}$ where t = tense. The verbal form has no gender; hence g is dropped; it has a tense; hence t is included.

Just as in mathematics, we raise the powers of a number by using exponents and write it as x^n, in linguistic analysis we raise the powers of the suffix and represent them in the above way. All these powers of the suffix are then automatically transferred to F and the F will assume the form as $N + S^{g+n+r+a+m} = F^{g+n+r+a+m}$(if nominal) or $N + S^{t+n+r+a+m} = F^{t+n+r+a+m}$(if verbal).

The most important non-formal category, from the point of view of our present study, is, however, the *vacana* or the number. The significance of *vacana* having a place in the speech itself in the form of a *pratyaya* lies in the fact that the people who used this device for indicating the number seem to have a keen mathematical consciousness or sense. The discussions on the meaning of number by the grammarian like Kauṇḍabhaṭṭa and the Naiyāyika like Praśastapāda bring out clearly the mathematical attitude of the ancient speakers of Sanskrit in applying the suffix for number also. Pāṇini has defined the *vacana* in the two *sūtras,* viz. *dvyekayor dvivacanaikavacane,* 1.4.22 (for singular and dual number) and *bahuṣu bahuvacanam,* 1.4.21 (for plural number).The necessity of including the number also in the suffix seems to have arisen from the fact that in nature certain things are found to exist in groups—of twos, threes, fours etc, although generally things exist in isolation or singularly. To take an example from Veda, the Aśvins, the *dyāva-pṛthivī* etc. are always found to be in a group of two; the Ṛbhus, the Maruts etc. are found to be in groups of more than two. Other deities like Agni, Vāyu, Indra etc. are found to be in singularity. The Vedic people or rather the pre-Vedic speakers of Sanskrit language, from whom the Vedic people inherited it, therefore, seem to have developed a number-sense which, they thought, must be included in the speech itself in the form of a suffix. And this mathematical consciousness on the part of the Vedic or pre-Vedic speakers of Sanskrit language is really an intellectual development and speaks for a sufficiently advanced state of civilisation in such ancient times as the Vedic. Not only this. But such a mathematical consciousness is not exhibited by any other ancient language contemporary to the Vedic one (if such a one exists or is available). Since the Vedas are the oldest literature no other language contemporary to the Vedic one is unfortunately available. But that the use of number-signifying suffixes with the words themselves indicates a sufficiently developed mathematical consciousness in the Vedic civilisation is certain. And conversely it is this number-consciousness which seems to have guided the ancient speakers of Sanskrit in using suffixes for the numbers. For them, therefore, the use of *puruṣau* (du. of *puruṣa*) is sufficient to convey the sense of 'two persons'; the usage *dvau puruṣau* from this point of view is unnecessary, the number-word *dvau* being redundant; so also in *ekaḥ puruṣaḥ,* the word *eka* is redudant. We, however, do find expressions like *dvā sakhāyā, trayaḥ keśinaḥ* etc. in the Veda, in which the number-words *dvā* (=*dvau*), trayaḥ etc. are used. But if we examine all such occurrences, we find that they use the number-words as numerical adjectives (*saṁkhyā-viśeṣaṇa*) in the case of those words which otherwise are generally found in other numbers also. We can find the word *sakhi* in sing. du. as well as plu. also as *sakhā sakhāyau* and *sakhāyaḥ*. So also with other cases. But the Veda never qualifies by number-words those things which are always found in groups of specific number. For example, we will never find the numerical adjective *dvau* in the case of Aśvins, *dyāvā-pṛthivī* or Mitrā-varuṇau; or the adjective *bahavaḥ* in the case of Ṛbhus of Maruts or Viśve devāḥ. The Vedic people thus seem to have a very clear idea, within their frame of knowledge, as to which things are found in specific numbers and which are not.

In the case of the former, they never use the numerical adjective; in the case of the latter, they use it.

If this line of thinking is correct, (and it is supported by facts from Veda also), it shows that the Vedic, or even pre-Vedic, people seem to have found out, after a long investigation and study, first, that certain things do not exist in any groups at all and certain other things stay always in groups. In the case of the latter also, they seem to have examined all the possible groups and found out the definite number of elements in the groups. They then seem to have classified the groups into groups of twos, groups of threes etc. Aśvins or Mitra-varuṇau, for example, are always found to exist in groups of twos, and so on. The later mathematicians seem to have taken a clue from this and started using the words themselves for the number; the *sūrya* (=the Sun), the *candra* (=the moon) etc. are found to exist always alone, that is, as one. The words *sūrya* (and its synomys), *candra* (and its synonyms) etc. came to be used later on as synonyms of the word *eka;* the Aśvins, always in two, came to be used for two and so on. Though this tendency to use the word-numerals as we may call them developed in later post-Vedic times, it seems to have a long mathematical back-ground and history. It must, however, be noted that the use of such word-numbers is not at all found in any of the nine Vedic saṁhitās taken here for study.

We thus find that there are two ways of signifying numbers in the Veda: (i) to use a number-word as numerical adjective qualifying the substantive as in *dvā sakhāyā,* or (ii) to apply a suffix invested with number to the word itself as in *aśvinau* or *marutaḥ* etc. Later on in post-Vedic literature, we find a third way of signifying numbers; and that is to use the word with fixed number for the number itself, as in *candra* = 1, *netra* = 2 etc. The only linguistic category which has no number is that of the indeclinables or *avyayas* including *nipātas.* Pāṇini has enumerated all the indeclinables in the *sūtra, svarādinipātam avyayam,* 1.1.37.

Though there are infinite numbers in mathematics, the suffixes in Sanskrit invested with the power of signifying numbers are only for three numbers, viz. *ekavacana* or singular, *dvivacana* or dual and *bahuvacana* or many i.e. more than two. The restriction to only three numbers is obviously because of the fact that we cannot have infinite suffixes for signifying infinite numbers.

Another peculiarity of the Veidc language which leads us to conjecture about the mathematical consciousness on the part of the Vedic people is the different Vedic metres, called *chandas.* "The main principle governing Vedic metre (the source of all later Indian versification) is measurement by number of syllables" (cf. *A A Macdonell, Vedic Grammar,* 1953, p 436). We have in the Veda chiefly the following five metres, viz. Gāyatrī of 24 syllables, the Anuṣṭubh of 32 syllables, the Triṣṭubh of 33 syllables, the Paṅkti of 40 syllables and Jagatī of 48 syllables. Besides, we have other varieties of metres, which are mixed, of 28 syllables like Uṣṇih, Pura-uṣṇih and Kakubh, of 36 syllables like Bṛhatī, of 40 syllables like Sato-Bṛhatī, of 60 syllables like Atiśakvarī, of 68 syllables like Atyaṣṭi; moreover we have what are called strophic stanzas like Pragātha, Kākubha Pragātha and Bārhata Pragātha. The composition of all these different types of metres presupposes a meticulous counting of the number of syllables in the Vedic stanzas. Not only this. But even the versification is not only mechanical; it has a kind of melody, rhythm and music which is still appealing today. Does all this mean that the Vedic people just composed the metres without any sense of counting and measuring? Did they not know computation?

II

Does the Veda really contain anything concerning mathematics which can be a subject for investigation? What does Veda really proclaim to stand for? or, what is the aim of Veda? These are some of the problems which one can ask while interpreting the Veda because the line of interpretation to be adopted depends very much on an answer to these problems. Let us try to get the answers to the above problems from the traditional point of view.

Traditionally speaking, the Veda has the following six disciplines as auxiliary means for its interpretation. They are: *śikṣā, kalpa, nirukta, chandas, vyākaraṇa and jyautiṣa.* Besides these six auxilaries called *Vedāṇgas,* we have the science of *pūrva mīmāṁsā* which also tries to interpret the Vedic passages. Of these six *vedāṅgas,* the *śikṣā, nirukta* and *vyākaraṇa* can be grouped under one category of linguistic interpretation and explanation of the Vedas. These three sciences deal with the different aspects of the Vedic language, viz. the phonetic aspect which is the scope of *śikṣā,* the etymological which is that of *nirukta* and the linguistic and grammatical which is the aim of *vyākaraṇa.* The Mimāṁsā referring to the *pūrva mīmāṁsā,* which tries to interpret the Veda from semantic and ritual point of view can also be grouped under this class. The *Kalpasūtras* attempt to stuyding in detail the procedure of sacrificial ritual as available in and based on Vedas. The Pūrva Mīmāṁsā also shares some part of this ritualistic interpretation. The science of *Chandas* (= 'metre') aim at analysing and describing the different metres in the Vedas. It can also thus be grouped along with the above three sciences *viz. Śikṣā, Nirukta* and *Vyākaraṇa* under the general title of 'sciences attempting a linguistic interpretation of the Vedas and meterical point of view.' All these sciences developed their own principles of interpretation, description and analysis of the Vedic language. The *Kalpasūtras* and Pūrva Mīmāṁsā have also their own way of analysis, description and interpretation of the Vedas. All these sciences emphasize only two aspects of the Veda, viz. the linguistic and the ritual. And none of them looks at Vedas from mathematical point of view.

The only auxiliary science which offers a possibility of studying Veda from mathematical point of view is that of *Jyautiṣa,* popularly written and pronounced as *Jyotiṣa.* It studies and describes the 27 constellations, their places in the sky and their properties. The earliest work in *Jyotiṣa* which helps the Vedic interpretation is the *Vedāṅga-Jyotiṣa.* The necessity for astronomical study for the Vedic interpretation arose out of the fact that the ritual and its details as are given in the Vedas are prescribed to be performed only on proper times, called *muhūrtas*; the *muhūrta* refers to a particular, fixed position or combination of the different planets in different constellations. It is, therefore, not possible to find out the *muhūrta* without studying the constellations and the planets. It is thus out of the necessity of finding out the *muhūrtas* for the sacrificial rituals that the science of astronomy was born and pursued. The following verse which is oft-quoted emphasizes the important status astronomy enjoys in Vedic interpretation:

vedā hi yajñārtham abhipravṛttāḥ
kālānupūrvā vihitāś ca yajñāḥ/

tasmād idam kālavidhānaśāstram
yo jyotiṣam veda sa veda vedān//

("The Vedas aim at sacrificial performances; these performances are stipulated to be performed on specific times; the one, therefore, who knows the science of time, viz. the *jyotiṣa* alone knows the Vedas").

The astronomical studies cannot in turn be pursued without mathematics. And thus at this stage of our Vedic study the Vedas offer a possibility of finding out at least some mathematical data in the Vedic texts. And some scholars like B.G. Tilak and H. Jocabi have actually studied Veda from astronomical point of view.

The Sanskrit language is said to contain a voluminous literature on Astronomy.[4] Even the oldest book on this subject viz. the *Vedāṅga Jyotiṣa* which describes the 27 constellations and studies the sky also in turn is either incomplete or seems to be an abridged form of a still earlier bigger work which is lost to us. Astronomy, however, cannot develop without the help of Mathematics. Though, therefore, mathematics in ancient India has always been a hand-made of Astronomy and hence can be said to occupy a secondary position with reference to the latter, it is just a matter of common sense to conclude that mathematics also must have been equally highly developed to help Astronomy. The actual mathematical and astronomical works which have come down to us, however, date far later than the Vedic *saṁhitās*. The astronomical and mathematical literature in between the periods of the Vedas and later works is unfortunately lost to us.

The *Vedāṅga-Jyotiṣa*, as the name indicates, sets out to be an auxiliary to the Vedic studies. It helps to know the astronomical data from the Veda. If, therefore, the Vedas contain any astronomical considerations, they should also contain some mathematical considerations. Unfortunately, no mathematical work as an auxiliary to the Vedic studies has come down to us under a title of Vedāṅga Gaṇita (like the Vedāṅga Jyotiṣa).

Again, whatever bulk of post-Vedic astronomical literature is said to contain mathematics contains more on Astronomy and very less on mathematics, obviously because of the fact that it is not a text on mathematics. The only perfect text on mathematics that has come down to us is that of *Leelāvatī* by Bhāskarācārya; and that too, very late–as late as the 11th century AD. In the absence of any data to that effect, we are unable to know the history of mathematics in ancient India. Even the *Śulbasūtras* which claim to be geometrical works are very much later than the Vedas.

Many books on Vedic mathematics have been written and published recently. The most recent one is by Jagadguru Shankarācārya Swāmī Śrī Bhāratī Kṛṣṇa Tīrthaji entitled '*Vedic Mathematics*'.[5] But after examination it is found not to be Vedic and hence the title sounds a misnomer. None of the works on Indian mathematics even touch the Vedic *saṁhitās*, not to say about exhaustive examination of the same. It is with the view to examine and ascertain whether the Vedas really contain any considerations on mathematics in the real sense that the present study has been undertaken. It is for the same purpose that the scope of the Vedic data on which to work is purposefully limited to the nine Vedic texts which go by the name *saṁhitās*. The Brāhmaṇa and upaniṣadic literature, which also is included under the title 'Vedic' traditionally, is, therefore, purposefully excluded. The purpose is to know the matiematical development in the times of strictly the Vedic *saṁhitās*. No study on Indian or Vedic mathematics uptill now has been based on what is strictly 'the Vedic', meaning thereby the Vedic saṁhitās proper.

The scope of the work is also limited to only the arithmetical part of mathematics which includes arithmetic, geometry, trigonometry and algebra. The geometrical and algebraical parts are excluded. If the geometrical and algebraical aspects were included, the work would have been too bulky.

As one goes through the Vedic *saṁhitās* and examines and interprets the mathematical data, one cannot but be impressed by the high stage of mathematical development in those ancient times. Not only this. But one has to pre-suppose a very long tradition of mathematical studies in pre-Vedic times also. Because whatever arithmetical in particular and mathematical in general conclusions are stated here in the present work cannot be the sudden off-shoot of those times only. They seem to have had a long tradition and history in the field. The conclusions will convince one of the high stage of not only Vedic mathematics but of Vedic civilisation itself. The data presented here will convince one that writing was known in the Vedic times. The concept of mathematical zero, the zero in the symbol 10 for *daśa*, the technical terms like *yajñena kalpantām* or *sarvasmai svāhā* , giving out the procedure to arrive at the concept of infinity, the different arithmetic and geometric series–all these facts show that the mathematical data in the hoary antiquity of the Vedic times cannot be a creation of a 'primitive' society; it can only come from a highly developed and one of the most advanced people and civilisation. The idea of a device of infusing the suffixal elements themselves with the power of signifying the number also, besides other non-formal elements like meaning, gender etc., cannot be taken to strike a primitive people who are said to roam the vast territories on the earth in search of the primary human necessities like food, clothes and a habitat to dwell in. Read in the context of the ancient times in which the Vedic civilisation flourished, the above achievements in the mathematical field cannot be underestimated. It should also be remembered that while not underestimating the mathematical development in the ancient times of the Vedas, the pendulum of thought should not be allowed to swing to the other extreme of over-estimation. Some people cherish very high ideas about the Vedas and about whatever Veda contains. The mathematics in the Veda from their point of view is something which even the modern civilisation has no idea of, which is totally different from the modern one and which is even higher than the modern mathematics. It might be so–we do not know. But judged from the empirical standards of interpretation, the Vedas give us the picture presented in the present work. The picture may disappoint some people because the mathematical picture presented here is very simple and is what they already know. True. But one should not forget that what they know is only because of Vedic mathematics and is based on the same, and that it is this Vedic system of mathematics that has been adopted and followed uptill now from the ancient Vedic times and that too throughout the present world. And this is no small achievement or contribution of the Vedas. What we can say at the most is that since the Vedas are the oldest extant literature of the world, which has come down to us in tact (thanks to the Vedic reciters!) it is likely that other non-Vedic civilisations might have borrowed the knowledge of mathematics from the Vedas or at least from pre-Vedic literature which is not available. The question who borrowed from whom can only be answered after a comparative study of the mathematical works of those civilisations.

III

The subject of Vedic mathematics was haunting me since the time I started doing Vedic research in 1954 under the guidance of the late Prof. Dr. S.S. Bhawe in the M.S. University of Baroda, Baroda. But I started the actual work on the subject only in 1975. I prepared a very short general research paper of about five pages for submitting it in the All India Oriental Conference which was held in Pune in 1978. And a five-line summary of the paper was published by the AIOC of 1978. When Dr. V.N. Jhā assumed the charge of the Director, CASS, Pune, in 1986, he insisted that I should present the Vedic mathematical data in the form of a monograph. And accordingly, I am putting my findings on the subject before the world of scholars both in mathematics as well as Sanskrit. It is left upto them to judge how far I have been successful in this endevour.

I have practically examined almost all the data. Even if some data are left out, that, I think, will not change the final conclusions. More data might perhaps only add to the bulk of the work.

I would be ungrateful if I do not express my indebtedness to all those who, out of labour of love, helped me in one way or the other in writing this book. First and foremost is Prof. Dr. V.N. Jha who encouraged me to write on this difficult topic. But for his friendly encouragement and insistence, I would have perhaps never written at all so exhaustively on this subject. Dr. *V.T. Zambare*, who retired from the University services recently also deserves my thanks. Even if retired, he also insisted and saw I complete the work. The next person to whom my thanks are due is Dr. Rayomond D. Doctor, Reader in French Language and Literature, of the Department of Modern European Languages, University of Poona, Pune. He helped me in providing the data of the number-words from the European languages viz. Greek, Latin, German, French, Italian, Spanish and Russian. The data being very important for a comparative study are published in the book as Appendix A. The data, I feel, have enhanced the value of the work. I am extremely indebted to Dr. T.T. Raghunathan of the Department of Mathematics, University of Poona, Pune, who provided me the literature, books and references from modern mathematics. He helped me in clearing many mathematical concepts. I have also to thank Dr. J.R. Joshi, of the Department of Sanskrit, University of Poona, for providing me the text of *Milindapañha* and examining the proofs. Last but not the least, the Research Associates from the CASS, Pune, viz. Dr. (Miss) Nirmala Kamat, Dr. (Mrs.) Kanchan Mande, Dr. (Mrs.) Anuradha Pujari, Dr. (Mrs.) Manik Thakar and Dr. (Mrs.) Bhāgyalata Pataskar also deserve my thanks; they read and re-read the manuscript of the work and suggested improvements in my writings. But for their suggestions, many of the arguments in the work might have been either not clear or been misunderstood. I am very much thankful to the Sadguru Press for the fine and efficient printing and the Proprietor of the Indian Books Centre, New Delhi, for publishing it. But for their interest in academic matters, the book would have taken a longer time for publication.

11th April, 1992 **M.D. Pandit**
Ramanavamī
Pune.

NOTES AND REFERENCES

1. Quoted by *J.R. Firth, The Tongues of Men and Speech* Oxford University Press, London, reprinted 1966, P. 99; for different types of syntaxes, *cf. ibid.* pp 77-83; *cf.* also, *C.F. Hocket, A Course in Modern Linguistics.* The Macmillan Comapny, New York, 1958, pp. 177-191; also pp. 209-220

2. *cf.* Patañjali on the Pāṇinian *sūtra,* 1.3.3: *ke punar vyavasitāḥ? dhātu-prātipadika-pratyaya-nipāta-āgama-ādeśāḥ.*

3. All these are non-formal categories. For the relation between the formal and non-formal categories in Pāṇini's grammar, *cf. M.D. Pandit,* 'Formal and Non-Formal in Pāṇini', ABORI, 1975

4. The CASS is going to publish shortly a full bibliography of Astrology and Astronomy; *cf.* also, David Pingree, *Jyotiḥśāstra,* A History of Indian Literature, Otto Harrasso witz, Wiesbaen, 1981.

5. The book is published by Hindu Vishva Vidyalaya Publication Board, BHU. Varanasi, 1965.

Contents

1

Introductory

The Vedic civilisation is one of the oldest civilisations of the world. The other known, ancient civilisations are the Babylonian on the banks of the rivers Euphrates and Tigris, the Egyptian on the banks of the river Nile, the Chinese on the banks of the river Yang Tse, the Indus Valley on the banks of the river Sindhu or Indus and the Mayan in South America. The Vedic civilisation flourished on the plains between the two great rivers of India, viz., the Gaṅgā and Sindhu. Except the Indus Valley civilisation, all other civilisations provide us sufficiently good data about themselves in the form of either archeological findings or literature or both which throws a considerable light on their character, type, form and extent. The Indus Valley civilisation, however, provides us only the data in the form of archeological findings of some seals which are as yet not deciphered. Of the remaining five, the Vedic civilisation strikes a difference. While the Babylonian, the Egyptian, the Chinese and the Mayan civilisations provide the data for study in the form of both the literature as well as archeological findings, the Vedic civilisation can be studied only through the literary data in the form of the Vedas; it does not provide us any archeological findings through excavations. Even whatever Vedic literature has come down to us has come down only through oral tradition and not in the form of any written documents. While studying the Vedic civilisation, we have to rely on the Vedic texts themselves since there are no other means like the archeological findings to testity the accuracy of the literary evidence provided by the Vedas.

The word *Veda*, from √*vid* 'to know', means 'knowledge'. The Vedas are thus taken to be the reservoir of the ancient knowledge gathered by the Aryan people in ancient, pre-Vedic times. And if it is the scientific knowledge which requires preservation and which, therefore, has been preserved with utmost care and sanctity throuh oral tradition throughout the long span of five thousand years, it must contain an element of precision. The desire for precision in knowledge requires an attitude for expressing scientific facts and theories in a quantitative way on the part of the scientists or investigators. Such an attitude in turn requires and presupposes the knowledge of mathematics which alone helps us in quantification. The knowledge of the basic, natural numbers from 0 to 9 is, therefore, an *ā priori* requirement for quantification, since mathematics starts with the procedure of computation and counting with numbers. If, therefore, the Vedas claim to contain the ancient, pre-Vedic knowledge, it automatically follows that they must also contain the knowledge of at least elementary mathematics involving numbers which has helped them in the quantification of their knowledge. In other words, if the Vedic authors have quantified all their knowledge which was worthy and capable of quantification, they must, at least in some places, be found to have been using mathematics and the natural numbers in their compositions.

As will be clear from the following pages, the Vedic authors do exhibit the knowledge of mathematics in their compositions. They have used the numbers, the four basic mathematical operations of addition, subtraction, multiplication and division; also they know to build up infinite number of infinite series with different progressions. All such data can be collected and be subjected to critical study.

2

The Scope of the Present Work

A number of views regarding the stage of Vedic civilisation have been expressed by different scholars, Indologists, Linguists, Historians, Archeologists etc.—from the one which assumes the Vedic culture as primitive, simple without any complexities and any knowledge of different sciences to the other which says that it was the most highly developed civilisation, comparable to the modern one, with discoveries and inventions in practically all fields of life. Each of the points of view supports its conclusions - rather, convictions, with passages from the Vedic literature.

Perhaps the most certain and accurate means to measure the development of a civilisation and culture is the study of mathematical knowledge as displayed in its literature. No civilisation can develop and progress without the development and progress in mathematics. Converely, progress in mathematics is a measure of the development of a civilisation.

The scope of the present monograph, therefore, is set out to be the study of the knowledge of mathematics of the Vedic people, because, as we have said before, we will not be able to know the real stage of the Vedic civilisation without knowing the stage of development in mathematics. The scope of the study is, therefore, in the first place, limited to the study of natural numbers, and the four, main, basic mathematical operations only. Though mathematics comprises of the three branches viz. arithmetic, algebra and geometry, the study is purposefully restricted to arithmetic only, because it is only with numbers that the real idea of quantification dawns on the mind.

Secondly, the Vedic literature really embraces a vast field: the Vedic *samhitās*, the Brāhmaṇas and the Upaniṣads. The study is restricted in this case only to the study of the nine main Vedic texts which go by the name of *saṁhitās* and which are really taken as *saṁhitās* by the Vedic recensions. They are as follows: the Ṛgveda *saṁhitā*, the *Vājasaneyi Śukla Yajurveda saṁhitā,* the Kāṇva *saṁhitā*, the Sāmaveda *saṁhitā,* the Atharvaveda *saṁhitā* (including both the recensions viz. the Śaunaka *saṁhitā* and the Paippatada *saṁhitā*), the Taittirīya *saṁhitā,* the Maitrāyaṇī *saṁhitā, the Kāṭhaka saṁhitā* and lastly the Kapiṣṭhala Kaṭha *saṁhitā.* There are no other texts which go by the name of, and which are taken to be Vedic *saṁhitās* as such.

The purpose in limiting the study only to the *saṁhitā*-texts is to know the oldest stage of the mathematical knowledge in ancient India. Generally, all the works on ancient Indian mathematics which have been written uptill now start at the earliest from the study of the *Śulbasūtras* which come definitely after the Vedic *saṁhitās*. Most of the works start with Āryabhaṭa I (Ist century A.D.) and do not touch even the Brāhmaṇa literature, let alone the Vedic literature. All the development in mathematics in the post-Vedic age, therefore, looks like a sudden off-shoot without any prior tradition in the field. It is with the view, first, to know the mathematical development in pre-*sulba-sūtra* and pre-Brāhmaṇa periods that the present study is restricted only to the *saṁhitā* texts. Secondly, such a study from the beginning of the Vedic period will certainly help us to build up step by step the authentic history of mathematics in ancient India.

The study is also restricted to the arithmetical part in mathematics. Not that the Vedic people did not have the knowledge of what we today call as geometry and algebra. But the inclusion of the algebraic and geometrical interpretation of the *saṁhitā*-verses would have made the work too bulky. Hence the geometrical and algebraic studies are also excluded from the present study; the arithmetical aspect is more concentrated upon.

An important point requires to be remembered in this connection. The Vedas are not the texts on mathematics. In general, we do not know the aim, purpose and scope of the Vedic *saṁhitās*. Whatever mathematical data are presented here for study are drawn by implication on the basis of the interpretation of the Vedic texts. Sometimes the Vedic texts are very clear and state in expticit words the results arrived at; *cf.* for example, the passage from TS. 7.4.11: *dvau ṣaḍahau bhavataḥ. tāni dvādaśāhāni sampadyante* (= two six-days make twelve days). In some cases, however, we have to know the mathematical results by implication; *cf.* for example the passage from MS. 3.9.3: *daśa vai paśoḥ prāṇāḥ; ātmā ekādaśaḥ* in which the result 10+1 = 11 is to be known by implication. But in general, even the implied mathematical statements and expressions are comparatively clear even to one untrained in mathematics.

Though the nine Vedic *saṁhitās* are not composed in the same place and time, they as *saṁhitā*-texts are taken as a whole literature because they form the basis of interpretation in the later brāhmaṇa literature.

We start with recording the cardinal number-words as stated in the *saṁhitā*-texts taken as a whole for our present study.

The following Table No. 1 gives the list of all the cardinal number-words actually recorded in the nine Vedic *saṁhitās*.

Table No. 1

Sr. No.	Number-words occurring in the saṁhitās	Modern numerical symbol indicating the value
1.	*eka*	1
2.	*dvi*	2
3.	*tri*	3
4.	*catur*	4
5.	*pañchan*	5
6.	*ṣaṣ*	6
7.	*saptan*	7
8.	*aṣṭan*	8
9.	*navan*	9
10.	*daśan*	$10 = 10^1$
11.	*ekādaśan*	11
12.	*dvādaśan*	12
13.	*trayodaśan*	13
14.	*caturdaśan*	14
15.	*pañcadaśan*	15
16.	*ṣoḍaśan*	16
17.	*sapadaśan*	17
18	*aṣṭādaśan*	18
19.	*ekonaviṁśati*	19
20.	*ekānnaviṁśati*	19
21.	*navadaśan*	19
22.	*vimśati*	20
23.	*ekaviṁśati*	21
24.	*dvāviṁśati*	22
25.	*trayoviṁśati*	23

26. caturviṁśati 24
27. *pañcaviṁśati* 25
28. *ṣaḍvīṁśati* 26
29. *sapta viṁśati* 27
30. *tri-nava* 27
31. *aṣṭāvīṁśati* 28
32. *navavīṁśati* 29
33. *ekānnatriṁśat* 29
34. *triṁśat* 30
35. *ekatriṁśat* 31
36. *dvātriṁśat* 32
37. *trayastriṁśat* 33
38. *catustriṁśat* 34
39. *pañcatriṁśat* 35
40. *ṣaṭtriṁśat* 36
41. *aṣṭātriṁśat* 38
42. *ekānnacatvāriṁśat* 39
43. *catvāriṁśat* 40
44. *ekacatvāriṁśat* 41
45. *catuś catvāriṁśat* 44
46. *pañca catvāriṁśat* 45
47. *navacatvāriṁśat* 49
48. *ekasmāt-na-pañcāśat* 49
49. *ekasyai-na-pañcāśat* 49
50. *pañcāśat* 50
51. *ekapañcāśat* 51
52. *dvāpañcāśat* 52
53. *tripañcāśat* 53
54. *pañcapañcāśat* 55
55. *ṣaṭpañcāśat* 56

56.	*aṣṭāpañcāśat*	58
57.	*ekānna ṣaṣṭi*	59
58.	*ṣaṣṭi*	60
59.	*ekaṣaṣṭi*	61
60.	*catuṣṣaṣṭi*	64
61.	*pañcaṣaṣṭi*	65
62.	*aṣṭāṣaṣṭi*	68
63.	*navaṣaṣṭi*	69
64.	*saptati*	70
65.	*ekasaptati*	71
66.	*dvāsaptati*	72
67.	*ṣaṭsaptati*	76
68.	*saptasaptati*	77
69.	*ekānnāśīti*	79
70.	*aśīti*	80
71.	*ekāśīti*	81
72.	*catur-aśīti*	84
73.	*pañcāśīti*	85
74.	*aṣṭāśīti*	88
75.	*navāśīti*	89
76.	*navati*	90
77.	*ekanavati*	91
78.	*dvānavati*	92
79.	*pañcanavati*	95
80.	*ṣaṭ-navati*	96
81.	*aṣṭānavati*	98
82.	*ekānnaśata*	99
83.	*śata/ekaśata/daśati*	100 or 10^2
84.	*dviśata*	200
85.	*triśata*	300

86.	*catuśśata*	400
87.	*pañcaśata*	500
88.	*daśaśata*	1000 or 10^3
89.	*sahasra*	1000 or 10^3
90.	*trisahasra*	3000 or 3 X 10^3
91.	*catuḥsahasra*	4000 or 4 X 10^3
92.	*ṣaṭsahasra*	6000 or 6 X 10^3
93.	*ayuta*	10,000 or 10^4
94.	*niyuta*	100,000 or 10^5
95.	*prayuta*	1000,000 or 10^6
96.	*arbuda*	10,000,000 or 10^7
97.	*nyarbuda*	100,000,000 or 10^8
98.	*samudra*	1000,000,000 or 10^9
99.	*madhya*	10,000,000,000 or 10^{10}
100.	*anta*	100,000,000,000 or 10^{11}
101.	*parārdha*	1000,000,000,000 or 10^{12}

Besides the numbers recorded above, we have again the following numbers in between the bases of one hundred and one thousand, and one thousand and ten thousands. These numbers, or rather number-words, have no single term but are calculated in terms of the multiples of ten, hundred and thousand. For the sake of avoiding unnecessarily long list of all the numbers recorded in the Saṁhitāas, the numbers recorded only in the RV are reproduced here, because they are sufficient to give us the idea of how the Vedic people were expanding the number-system and communicated. They are follows:

śatam ekam ca (101), *śataṁ sapta ca* (100+7 = 107), *tisraḥ pañcāśataḥ* (50+50+50 = 150), *triḥ ṣaṣṭiḥ* (60+60+60 = 180), *dve śate* (100+100 = 200), *tisṛṇām saptatīnām* (70+70+70 = 210), *trīṇi śatāni* (100+100+100 = 300), *sākam triśatā ṣaṣṭiḥ* (60+300 = 360), *pañca śatā* (100+100+100+100+100 = 500), *sapta śatāni viṁśatiḥ ca* (700+20 = 720), *navatim nāvyā* or *navānām navatīnām* (9 ninties = 9 X 90 = 810), *sahasra* or *daśa śatā* (10 hundred = 1000), *triḥ sapta saptatīnām* (3X7X70 = 1470), *viṁśatim śatā* (20 hundred = 2000), *triṁśat śatam* (3000), *triṇi śatā trī sahasrāṇi triṁśatam ca nava ca* (300+3000+30+9 = 3339), *catuḥsahasram* or *catvāri sahasrā* (4000), *ṣaṣṭiḥ śatā* (6000), *aṣṭā sahasrā* (8000), *daśa sahasrā* (10,000), *ṣaṣṭiḥ śatā... ṣaṭ sahasrā ṣaṣṭiḥ ṣaṭ* (6000+6000+60+6 = 12066), *triṁśatam... sahasrāṇi* (30,000), *pañcāśat sahasrā* (50,000), *ṣaṣṭiḥ sahasram* (60,000), *ṣaṣṭiṁ sahasrā navatim nava* (60,099), and finally sahasrāṇi śatā (100,000) which is the highest number mentioned in the Ṛgveda. We have still one more number which may be interpreted as higher than the highest mentioned above viz. 100,000; the wording, which is ambiguous, is: *sahasrāṇi śatā daśa* which means differently if interpreted in various ways. If interpreted as *sahasrāṇi śatā +daśa*, it means 100,000+10 = 100,010; if interpreted as *daśa śatā sahasrāṇi* , meaning ten hundred thousands, it gives us the number, 1000,000, which will be the highest number mentioned in the Ṛgveda. There is still another number mentioned in the Ṛgveda; but it is not given in terms of multiples of ten, hundred or thousand; it has an independent, single term. The number is *ayuta* whose value cannot be exactly determined. If later mathematical or astronomical literature is to be believed, *ayuta* represents the number *daśa sahasrā* i.e. 10,000 and the expression *catvāri ayutā* in RV. 8.2.41 would mean 40,000. Sāyaṇa at RV. 4.26.7 renders *ayuta* by *ayutasaṁkhyākam aparimitasaṁkhyākam ity arthaḥ* , which shows that even Sāyaṇa is not sure of the exact value of the number signified by the term *ayuta*. It is to be noted that Ṛgveda does not define any of the numbers. It is also to be noted that there is no explicit mention of the number zero in the Ṛgveda. Also, there is absolutely no mention of the negative numbers. These numbers noted above are expressed not by single words, but by phrases containing number-words. These phrases, it should be incidentally noted, state down in a way the process of addition.

VS. 17.2 mentions a series of numbers starting from the basic 10, each succeeding one of which is ten times the immediately proceding one. Thus, starting from 10, we have, 10 (*daśa*), 100 (*śata*), 1000 (*sahasra*), 10,000 (*ayuta*), 100,000 (*niyuta*), 1000,000 (*prayuta*), 10000,000 (*arbuda*), 100,000,000 (*ny-arbuda*), 1000,000,000 (*samudra*), 1000,000,000,0 (*madhya*), 100,000,000,000 (*anta*) and lastly 1000,000,000,000 (*parārdha*).

The TS. (*cf.* 4,4.11.4) follows the VS. *ad verbatim*; but the KS (*cf.* 17.10) interchanges the places of *prayuta* and *niyuta*; thus, according to KS, VS. *niyuta* (10^5) = KS. *prayuta* (10^6); and VS. *prayuta* (10^6) = KS. *nīyuta* (10^5). The PVB (=pañca-viṁśa-brāhmaṇa) gives different names to the numbers; *cf. VS. arbuda* i.e. 10^7 = PVB. *nikharva* - 10^8, *Vāḍara* = 10^9, *akṣiti* = 10^{10} and *parārdha* = 10^{11}. The MS. follows VS upto the number *ayuta* i.e. 10^4; but lists and names the next numbers as *prayuta* - 10^5 (= VS. *niyuta*), *ny-arbuda* = 10^6 (= VS. *prayuta*), *samudra* = 10^7 (=VS. *arbuda*), *madhya* = 10^8 (= VS *nyarbuda*), *anta* = 10^9 (= VS. *samudra*) and lastly *parārdha* = 10^{10} (= VS. *madhya*).

It will be evident that all these numbers are multiples of ten and can be arrived at by raising the power or index of the number 10. Though, therefore, none of the Vedas states and names the numbers after *parārdha* i.e. 10^{12}, we can obtain the next multiples by raising further the powers of 10 and go on counting the numbers as 10^{13}, 10^{14}, 10^{15} etc., etc. and we can have infinite series like this. Though the VS. and other *Saṁhitas* do not lay down explicitly the procedure of raising the powers, the whole procedure is implied in the explicit statements of the series itself.

3

The Sequence or the Serial Order of the Numbers

The sequence of the integers from 1 to 10 is the same as is followed by us even to-day. In other words, the sequence of numbers has not changed at all throughout these ages.

RV. 6.45.5 (*tvam ekasya vṛtrahan avitā dvayor asi*) gives us that the number two follows the number one. The Ṛgvedic passages 8.45.34 (*mā na ekasminn āgasi, mā dvayor uta triṣu*), 10.48.7 (*abhīd ekam eko asmi niṣṣāḍ, abhi dvā, kim u trayaḥ karanti*) state the sequence of the first three numbers as 1, 2, 3. The *ṛc* 1.155.5 (*dve idasya kramaṇe......tṛtīyam asya nakir ā dadharṣati*) states that 3 follows 2. The passage, *guhā trīṇi nihitā neṅgayanti turīyaṁ vaco manuṣyā vadanti* , RV. 1.164.44 gives the sequence of 3 and 4, the latter following the former. The passage, *jyeṣṭha āha camasā dvā kareti.... trīn kṛṇavāma....caturas kareti,* RV. 4.33.5, gives the sequence of 2, 3 and 4 as the same as is followed to-day. That the number 5 follows the number 4 is given in the Ṛgvedic passage, 10.13.3: *pañca padāni rupo anvaroham catuṣpadīm anvemī vratena*, which means that once one follows by sequential order the number 4, he can reach (lit. ride) the number 5.

The ṛc 10.27.15 (*sapta vīrāso adharād udāyan, aṣṭa uttarāttāt .... nava paścāttāt....āyan, daśa prāk sānu virājati*) gives the order of the numbers from 7 to 10 as 7, 8, 9, 10.

That 11 is stated to come after the number 10 is clear from the Ṛgvedic passage, 10.45.85 (*daśāsyām putrān ādhehi, patim ekādaśam kṛdhi*).

The word *ekādaśa* used for the number 11 makes it clear that 11 = 1+10. This shows that the first series of single-digit numbers is over and the second series has started. It also makes it clear that the second series arises out of the first series by additing the numbers 1, 2, 3.....9 to the number 10. Thus, we have *ekādaśa* (*eka+daśa*) = 1+10; *dvādaśa* (= *dvā* i.e. *dvi*+daśa) = 2+10; *trayodaśa* (= *trayaḥ* from *tri+daśa*) = 3+10 and so on. This second series continues upto the number 19, which is represented as the sum of 9+10 by using the number-word *nava-daśa* (9+10) available in the VS.

The numbers 5, 6 and 7 are, nowhere explicitly stated to be in the serial order at least in the RV. Yet that they are the consecutive numbers coming after the number 4 can be guessed by the words *caturdaśa* (4+10), *pañca-daśa* (5+10), *ṣoḍaśa* (= *ṣaṭ+daśa* i.e. 6+10) and *sapta-daśa* (7+10). That the number *pañca-daśa* comes after the number *catur-daśa* is implicit in the Ṛgvedic statement, 10.114.7 & 8; *cf.* RV. 10.114.7: *caturdaśānye mahimāno asya* which is immediately followed by RV. 10. 114.8: *sahasradhā pañcadaśāya ukthā*. From the fact that 14 = 4+10, 15 = 5+10, 16 = 6+10 and 17 = 7+10, we can infer that though the serial order of the number 4, 5, 6 & 7 is nowhere explicitly given in the RV, it is implied in the above expressions.

The hymns nos. 5.15 and 5.16 from the AV, however, explicitly state the numbers from 1 to 10 in the serial order as: *eka, dvi, tri, catur, pañca, ṣaṭ, sapta, aṣṭa, nava, daśa* and *ekādaśa*. The rest of the numbers can then automatically follow.

VS. 9.31-34 also lists the first 17 numbers in the regular sequence which gives us that the number 5, 6, 7, 8, 9 and 10 follow the number 4 in that order; *cf.*

VS. 9. 31-34: *ekākṣareṇa prāṇam...dvyakṣareṇa dvipadaḥ... tryakṣareṇa trīn lokān...caturakṣareṇa catuṣpadaḥ... pañcākṣareṇa pañca diśaḥ...ṣaḍakṣareṇa ṣaḍṛtūn...saptākṣareṇa sapta grāmyān... aṣṭākṣareṇa gāyatrīm... navākṣareṇa trivṛtam... daśākṣareṇa virājam... ekādaśākṣareṇa triṣṭubham... dvādaśākṣareṇa jagatīm... trayodaśākṣareṇa trayodaśaṁ stomam caturdaśākṣareṇa caturdaśam stomam... paṇcadaśā-kṣareṇa pañcadaśam... ṣoḍaśākṣareṇa ṣoḍaśam... saptadaśākṣareṇa saptadaśam stomam udajayat,* which lists numbers 1-17 in a serial order. VS. 14.23 goes on to specify numbers 17-25 in the serial order; *cf. pañcadaśaḥ... saptadaśaḥ... aṣṭādaśaḥ navadaśaḥ... saviṁśaḥ... ekaviṁśaḥ... dvāviṁśaḥ... trayoviṁśaḥ... caturviṁśaḥ... pañcaviṁśaḥ.* After this, the series is not in the order; yet the counting continues upto the number 48; *cf. ekatriṃśaḥ... trayastriṁśaḥ... catustriṁśaḥ... ṣaṭtriṁśaḥ... aṣṭācatvāriṁsaḥ*. It is to be noted that for the number 19, we have two words, viz., *navadaśa* and *ekona viṁśati*. As is evident from the word *navadaśa* which can be split up as *nava+daśa* (i.e. 9+10), the author seems to start with the base 10 and arrives at the number 19 by adding successively the numbers from 1 to 9. The word *navadaśa* thus implies the process of addition. This word is invariably used in the VS. The other word viz. *ekonaviṁśati*, split up as *eka+ūna+viṁśati* (lit. 'one less than 20') shows the author is viewing the number 19 as 'one less than 20'. This word, therefore, implies the process of subtraction; thus 19 = 20-1. This word is found to be used only in the Atharvaveda for the first time. After the Atharvavedic stage, the word *ekonaviṁśati* replaces the word *navadaśa* which becomes obsolete; *cf.* AV. 19.23.16: *ekonaviṁśatiḥ svāhā.*

The number 20, expressed by the word *viṁśati* marks the end of the second series from 11-19 and the beginning of the next, third series from 21 onwards. The numbers in these series also are expressed as an addition of the numbers from 1-9 with the number 20. It is really interesting to note that the VS. 14.31 (*navaviṁśatyā stuvata*) expresses the number 29 also as *nava+* viṁśati (i.e. 9+20) and not as *eka-ūna-triṁśat* (i.e. 'one less than 30'). It seems, therefore, that primarily the development of numbers and the series seems to be based on the principle of addition of 1 to the previous number. This was in the times of the RV. and VS. The view of a number, like, say, 19 or 29, as 'less than the next by one' seems to have developed later on. This started from the times of the AV. onwards.

VS. 25.4,5 enumerate the numbers 3-13 in the serial sequence. VS. 39.6 lists the sequence of numbers 1-12. VS. 27.43 states in sequential order of the numbers 1-4. RV. and VS. 12.75 mentions the number 107 as an addition of 100+7; *cf. śatam dhāmāni sapta ca*.

The word *triṁśat* refers to 30; *catvāriṁśat* to 40; *pañcāśat* to 50; *ṣaṣṭi* to 60; *saptati* to 70; *aśīti* to 80 and *navati* to 90. For 100, the word used is *śata*; *sahasra* signifies 1000.

The RV. contains the words like *ayuta, prayuta, arbuda, anta, samudra,* etc. Yet, except the word *ayuta*, other words do not signify the numerical values.

The VS, as we have seen, contains a regular series from 1 to 10^{12}.

The number *śatam* marks the end of the two-digit series and the beginning of the three-digit series. The number *sahasra* marks the end of the three-digit series and the beginning of the four-digit series.

It will be clear from the above that for all practical purposes of general counting in every day life, the number 10 (*daśa*) seems to have been assumed as the primary base in Vedic times. It is also to be noted that each succeeding number is arrived at by adding the number 1 (*eka*) to the preceding number. The natural integers are thus clearly an example of what modern mathematics calls as 'the arithmetic progression with a difference of 1 between any two consecutive numbers'. Thus we have,

x, $x+1$, $x+1+1$, $x+1+1+1$ etc. as the general formula for the natural integers.

4

Characteristic Features of the Vedic Number-System

By filling up the unrecorded numbers in the Vedic number-system, we find that it gives a full-fledged number-system which we are following to-day. Even through large expanse of time and space, the Vedic system seems to have been adopted and followed by us *in toto* in its original form as given in the Vedas. As such the modern number-system inherits all the characteristic features of the old Vedic number-system. It is also true that in the process of adopting and following the old Vedic number-system, we have not made, even after or through a period of 4000-5000 years, any noticable contribution either by way of change or modification to the old system. The Vedic number-system exhibits the following chief characteristics:

4.1. Arithmetic Progression

Each succeeding number can be obtained by the addition of 'one' (1) to the immediately preceding one[1]. Thus 2 can be obtained by adding 1 to 1; 3 by adding 1 to 2 etc. There is, therefore, always a difference of 1 between any two consecutive numbers. In the words of modern mathematical terminology, the series of numbers increases in 'arithmetic progrogsion'; the whole series therefore, seems to have been based on the principle of 'arithmetic progression'.

4.2. 'Ten' as the radix

If we examine the number-words used for numbers after 'ten', we find that they contain 'two words'. For example, the word *ekādaśa* contains the word *eka* and *daśa* for the single number 'elevan'; so also with all the following numbers. The numbers from 'one' to 'ten' are, however, expressed only with one word. This means that the number 'ten' seems to have been used as 'the radix' or 'the base' for all the following numbers. With the number 'ten' (*daśa,* 10) as the radix, the Vedic number-system has turned to be 'a decimal system'.

The most important result of the base 'ten' has been that the series, which is one-digit series upto the number 'nine' changes its form to a two-digit series, in which the preceding number-member is 'one'. This continues upto the number 'nineteen' at which stage, the series again changes its form with the preceding number-member being 'two' standing for 'twenty'. The number 'twenty' is, as we know, a multiple of 'ten'. At thirty, the decimal place is occupied by the number 'three' standing for 'thirty'. In every successive series afterwards, the decimal place is occupied by numbers 'four', 'five' etc. in order, until we reach the number 'ninty nine'. At this stage, the series changes its level or rank to 'a three-digit' system and so on.

The base 'ten' thus is very important and has given to the system an absolutely systematic character with an element of predictability. The following explanation will clarify the above statement. We can see that the single-digit series from one to nine (1-9) has nine members in it. The next two-digit series from 10-99 has ninety members in it, which number is ten times that in the former series of single-digit. The next three-digit series from 100-999 has nine hundred members in it, which is ten times that in the preceding two-digit series and so on. To put it symbolically in modern notation, we have the following picture:

Single-digit series = 9 members = 9×10^0

two-digit series = 90 members = 9×10^1

three-digit series = 900 members = 9×10^2

four-digit series = 9000 members = 9×10^3 and so on.

From this we can easily predict the number of members in any-number-of-digit-series. Thus the number of members in a series (based on the number of digits) can be predicted in terms of the power of the base 'ten (10)'.

That the whole system is based on the basis of ten (10) is clear by the fact that the transformation of the single-digit number-structure into multi-digit number-structure can be easily brought about by multiplying the previous one by ten; thus, 1X10 = 10; 10X10 = 100; 100X10 = 1000 and so on. The commentators Uvaṭa and Mahīdhara on the VS. note this point specifically; *cf. Uvaṭa* on VS. 17.2: *evam ekāprabhṛti daśasaṁkhyāguṇitam parārdha-paryantam pūrvottarasaṁkhyā-samuccitam vardhamāna saṁkhyeyaniṣṭham saṁkhyājātam* . *cf.* also Mahīdhara on VS. 17.2: *atra ekādiparārdhaparyantaiḥ śabdaiḥ uttarotaram daśa-daśa-guṇitā saṁkhyā ucyate; ekā ekatvasaṁkhyāviśiṣṭā sā daśaguṇitā daśasaṁkhyām āpadyate; sā (= daśasaṁkhyā) daśaguṇitā śataṁ bhavati;... śataṁ daśaguṇitam sahasraṁ bhavati; sahasraṁ daśaguṇitaṁ ayutaṁ bhavati; ayutam daśaguṇitam niyutam bhavati....niyutaṁ daśaguṇitaṁ prayutaṁ bhavati....prayutaṁ daśaguṇaṁ koṭiḥ; koṭir daśaguṇa arbudam; arbudam daśaguṇam nyarbudam; nyarbudaśabdena abja-saṁkhyā jñeyā...tena abjam daśaguṇam kharvam, kharvam daśaguṇam nikharvam; nikharvam daśaguṇaṁ mahāpadmam; mahāpadmaṁ daśaguṇam śaṁkuḥ; śaṁkur daśaguṇaḥ samudraḥ; samudro daśaguṇaḥ madhyam; madhyam daśaguṇam antaḥ; antaḥ daśaguṇaḥ parārdhaḥ.*

To put the above explanation of VS. 17.2 by the commentator Mahīdhara into number-symbols, the following numbers are recorded by VS:

1.	*eka*	$1 = 10^0$
2.	*daśā*	$10 = 10^1$
3.	*śata*	$100 = 10^2$
4.	*sahasra*	$1000 = 10^3$
5.	*ayutam*	$10,000 = 10^4$
6.	*niyutam* or *lakṣa*	$100,000 = 10^5$
7.	*prayutam*	$1,000,000 = 10^6$
8.	*arbuda* or *koṭi*	$10,000,000 = 10^7$
9.	*nyarbudam* or *abja*	$100,000,000 = 10^8$
10.	*kharva*	$1,000,000,000 = 10^9$
11.	*nikharva*	$10,000,000,000 = 10^{10}$
12.	*mahāpadma*	$100,000,000,000 = 10^{11}$
13.	*śaṁku*	$1,000,000,000,000 = 10^{12}$
14.	*samudra*	$10,000,000,000,000 = 10^{13}$
15.	*madhya*	$100,000,000,000,000 = 10^{14}$
16.	*anta*	$1,000,000,000,000,000 = 10^{15}$
17.	*parārdha*	$10,000,000,000,000,000 = 10^{16}$

The VS. 17.2, however, does not note the words from 10-13, called *kharva, nikharva, mahāpadma* and *śaṁku*. The total number of numbers noted by the Veda, therefore comes only to 13. Without the commentary of Mahīdhara, the following will be the number-symbols of the numbers from 14-17:

samudra = 10^9; *madhya* = 10^{10}; *anta* = 10^{11} and *parārdha* = 10^{12}.

These are in all 17 numbers noted by Mahīdhara; how he says that the total numbers noted are 18 is not clear; cf. Mahīdhara =

on VS. 17.2: *evam ekādi-aṣṭādaśa-saṁkhyāsaṁjñāsammitāḥ.* The VS., however, notes only the 13 numbers from 10^0 to 10^{12}. The number 10^{12}, therefore, seems to be the highest number recorded by Vedic saṁhitās. It is to be noted that VS. 17.2 = KS 17.10 = KKS. 26.9.

4.3. Simple and Compound Numbers

When we go through the above list of cardinal number-words mentioned in the Vedic Saṁhitās, we find, grammatically speaking, two types of number-words. The one type of words from *eka* to *daśa* contains what in Pāṇinian terminology is called as one 'prātipadika'.[2] The other type of words from *'ekā-das'a'* onwards consists of two nominal bases or *prātipadikas*. The first member of the second type is always one of the words from '*eka*' to 'nava'; the other member is either 'das'a' or its multiples i.e. *daśa*, *viṁśati, trīṁśat, catvāriṁśat, pañcāśat, ṣaṣṭi, saptati, aśīti* or *navati.* The numbers above *śata* i.e. one hundred are not expressed in one single term or *prātipadika* consisting of either one or two nominal bases, but are expressed by resorting to the procedural way of addition; thus 'one hundred and one' is expressed as '*ekaṁ ca śatam ca*', (RV.1.117.18) 'One hundred and seven' is expressed as '*śatam sapta ca* ' (RV. 10.97.1) and so on.

If, suppose, we represent the *prātipadikas* or nominal bases by some symbol, say, N (meaning Nucleus)[3], we get that the number-word *eka* can be represented only by one symbol, viz. a single N, while the number-word, *ekādaśa, dvādaśa* etc. which are above *daśa* require two symbols, viz. N_1N_2 or N_1+N_2 i.e. two Ns as the latter words contain two nominal bases. Thus, *eka* = N, while *ekādaśa* = N.N.

In the terminology of Pāṇini's grammar, when two *prātipadikas* combine together to form a third *prātipadika*, the latter thus formed newly, is termed as the '*samāsa*' or compound.[4] Grammatically speaking, therefore, the number-words from *eka* or *daśa* are 'non-compounded *prātipadikas,*' while the number-words above *daśa* are 'compounded *prātipadikas*' or nominal word-structures. To transfer the same terminology to the number-symbols, we may say that the number-symbols or, in short numbers, from 1 to 9 are non-compounded numbers while those after 10 are compound numbers[5]. Yet, it must be remembered at this stage that the numbers after 'ten' are 'compound numbers' because the number-words for numbers after 'ten' are 'compound words'. It also needs to be noticed that though the number 'ten' in symbols as '10' looks a compound symbol and hence a compound number, since it consists of two digits, the number-word for 'ten', viz. *daśa* is not a compound word-structure according to the rules of Sanskrit grammar of Pāṇini, since it does not contain two words. Why then the number 'ten', 10 in symbols, is represented by two separate symbols–one for 'one' and the other for 'zero', instead of by one, single symbol?

If we examine closely the Vedic number-words, we find that while the numbers from *ekādaśa* onwards are expressed in terms of the numbers from *eka* to *nava*, the latter are not expressed in any terms at all. In other words, while the former require the help of the latter, the latter do not require any help at all. Put it in a different way, while the words or numbers from *eka* to *daśa* are underived, the words or numbers from *ekādaśa* onwards are derived from the former. To illustrate, *eka* is just *eka*; but *ekādaśa* = *eka+daśa*. Thus, the number-word, and consequently the number, *ekādaṣa* seems to have been coined in terms of *eka* and *das'a*. As a corollary of the above statement, we can say that the number *ekādaśa* is obtained by the addition of *eka* with *daśa*. This is clear by the following Vedic passages: KS. 26.4: *daśa vai paśoḥ prāṇāḥ ātmā ekādaśaḥ,* which is repeated in MS 3.7.7; 9.3; KKS. 41.2. KS. 29.9: *daśa vai puruṣe prāṇāḥ ātmā ekādaśaḥ* , KS. 28.3 explains the same thing in another context; *cf.* KS. 28.3: *daśa vasavaḥ indra ekādaśaḥ*; *daśa rudrāḥ indra ekādaśaḥ; daśa ādityāḥ indra ekādaśaḥ.* AV.5.15.1 only mentions the summation-procedure for 11; cf *ekā ca me daśa ca me.* So also, other numbers after elevan are expressed in terms of ten; *cf.* KS.33.2. *daśa vai puruṣe prāṇāḥ stanau dvādaśau,* in which 12 = 10+2; TS. 7.3.7.4 explains 15 as a sum of 10+5; *cf. pañcadaśa etāḥ; tāsāṁ yāḥ daśa...yāḥ pañca.* So also 14 = 10+4, for which *cf.* TS. 7.3.5.3. *caturdaśa etāḥ; tāsām yāḥ daśa...yāś ca catasraḥ diśaḥ.* By continuing the process of adding the numbers 1-9 to the number 10, they represent 20 as 10+10; *cf.* TS. 7.3.7.4: *viṁśo vai puruṣaḥ daśa hastyā aṅgulayo daśa padyāḥ.* Taking the bases 20, 30, 40, 50, 60, 70, 80 and 90, the AV. 5.152-9 explains 22 as 20+2 (*cf.* 5.15.2: *dve ca me viṁśatiś ca me*), 33 as 30+3 (5.15.3: *tisraś ca me triṁśac ca me*), 44 as 40+4 (5.15.4: *catasraś ca me catvāriṁśacca me*), 55 as 50+5 (5.15.5: *pañca ca me pañcāśac ca me*), 66 as 60+6 (5.15.6: *ṣaṭ ca me ṣaṣṭiś ca me*), 77 as 70+7 (5.15.7: *sapta ca me saptatiś ca me*), 88 as 80+8 (5.15.8: *aṣṭa ca me aśītiś ca me*) and 99 as 90+9 (5.15.9: *nava ca me navatiś ca me*).

We, therefore, find that taking clue from the procedure of addition of 1+10, or 10+1, the number-system in the Veda has been built up and also extended to count numbers beyond ten; thus 10+1 = 11; 10+2 = 12 and so on until 20 which is the next base for a change of series in the second rank; we can then proceed by adding 1 with 20 etc; and we get 20+1 = 21, 20+2 = 22 etc.

The same operation is repeated when we arrive at the three-digit number 100; and we build up 100+1 = 101, 100+2 = 102 and so on. We can thus build up an un-ending series and ranks by this method. That the same operation of addition of numbers 1-9 is to be repeated even after the three-digit number 100 (and also four-digit number of 1000 and so on) is clear from the Vedic passages in which the numbers after 100 etc. are recorded; *cf.*, for example, the numbers 101 as 100+1 (RV.1.117.18: *śatam ekaṁ ca meṣān*), 110 as 100+10 (RV.2.13.9 *śataṁ vā yasya daśa sākam adya*), 360 as 300+60 (RV.1.164.48: *sākaṁ triśatā...ṣaṣṭiḥ*).

The Vedic number-system is based on the gradation of ten. All other languages from the Indo-European, or rather Indo-Germanic family have followed the Vedic system of number-building. The following few examples for numbers 11-19 from some of the European languages will illustrate the point (*cf.* also Appendix A).

Table No. 2

Nos.Sanskrit	*Latin*	*Italian*	*French*
10. *daśa*	*decem*	*dici*	*dix*
11. *ekādaśa*	*un-decim*	*un-dici*	*on-ze*
12. *dvā-daśa*	*duo-decim*	*do-dici*	*dou-ze*
13. *trayo-daśa*	*tre-decim*	*tre-dici*	*trei-ze*
14. *catur-daśa*	*quattuor-decim*	*quattor-dici*	*quator-ze*
15. *pañca-daśa*	*quin-decim*	*quin-dici*	*quin-ze*
16. *ṣo-ḍaśa*	*se-decim*	*se-dici*	*sei-ze*
17. *sapta-daśa*	*septen-decim*	*dicīa-sette*	*dix-sept*
18. *aṣṭādaśa*	*octo-decim*	*dicī-otto*	*dix-huit*
19. *nava-daśa*	—	*dicia-nove*	*dix-neuf*
or			
19. *ekonaviṁśati*	*unde-vigīnti*	—	—

The languages other than Sanskrit and Latin as is clear from the above Table No. 2 change their number-counting a little from the numbers 17-19, by putting the word for 'ten' first, while Sanskrit and Latin are throughout consistent. Thus, for Sanskrit, the numbers from 11-19 are additions of 10 with numbers from 1 to 9. For Latin 19 is not 9+10 but 20-1. For other languages, the number 17, 18 and 19 are additions of numbers 7, 8 and 9 with the number 10 as is clear from the change of position of the word *dici* or *dix* from posterior to prior to the word signifying the numbers 7, 8 and 9. Even in the middle Indo-Aryan stage, viz. the Śauraseni, Mahārāshtrī, Ardhamāgadhī and Apabharaṁśa, the number-words from 1-9 occupy the first position, prior to the number-word *daśa*. (*cf.* the Appendices).

4.4. Grammatical Derivation of some numbers-words

As we have said above, though the compound number-words from *ekādaśa* onwards can be derived or dissolved, since they are compounds both from strictly grammatical and mathematical point of view, the single-digit numbers from *eka* to *daśa, viṁśati, triṁśat, catvāriṁśat, pañcāśat, ṣaṣṭi, saptati, aśīti, navati, śata* and *sahasra* are not derived by any Sanskrit grammarians. Yet, the *Uṇādi-sūtras*, following the etymological principles different from Pāṇini's have tried to derive mechanically the following five number-words, viz. *eka, tri, pañca, sapta* and *aṣṭa*. In doing so, the author of the *uṇādi-sūtras* seems to follow Yāska whose dictum in etymological consideration is: *na tv eva na nirbrūyāt* (= one should never say that he cannot derive even a single word)[6].

4.4.1. *The word eka*

The *uṇādisūtra*, 4526, viz. *iṇ-bhī-kā-pā-śaly-ati-marci-bhyaḥ kan*, lays down the suffix *kan* i.e. *ka* for the root √*i* 'to go' (2nd conj. Par.); thus,

i+ka

= *e+ka* (*guṇa* of *i* into *e*)

= *eka* (= one).

The book *Auṇādikapadārṇava*,[7] 3.103, gives three meanings of *eka* as 'one, the other and only'; *cf. eko'nyārthe pradhāne ca prathame kevale tathā*; following the above, BD. gives the three meanings as *mukhya, anya* and *kevala; cf.* BD. *mukhyānyakevalāḥ*.

4.4.2. *The word tri*

In the *uṇādisūtra*, 4950, *tarater ḍriḥ*, the word *tri is derived from the root* √*tṛ* 'to swim, to float'; the suffix is *ri* (*uṇādi, ḍri*), the process is:

tṛ+ḍri

= *tṛ+ri* (*ḍ* = 0 according to Pāṇini 1.3.7)

= *t+ri* (*ṛ* = 0 according to Pāṇini 6.4.143)

= *tri* (= three).

The *Auṇādikapadārṇava* does not note either the *sūtra* or even the form *tri.*

4.4.3. *The word pañca*

The *Auṇādikapadārṇava* (*ibid.* 1.820) does not derive this form grammatically. He, however, notes that the number-word *pañca* signifies a number between *catur* (i.e. four) and *ṣaṭ* (i.e. six); *cf. pañceti ṣaṭcaturmadhya-saṁkhyāvācīti dṛśyate; pañca,* therefore, denotes the number 'five'.

4.4.4. *The word sapta*

It is derived from the root √*sap* 'to gather, include' etc. (*cf. Pāṇinian dhātupāṭha, ṣapa samavāye*) with the suffix - *an*, which is nowhere given in any of the *uṇādisūtras.* What the *uṇādisūtra* states is only an *āgama* viz. *tuṭ* i.e. *t* before the suffix a. The *uṇādisūtra* which states the *āgama t* is: *sapy-aśūbhyām tuṭ ca,* 4358. The process is:

sap+an

= *sap+tuṭ+an*

= *sap+t+an* (*t* and *u* both are zeroed)

= saptan (= seven).

4.4.5. *The word aṣṭa*

The word *aṣṭa* is derived from √*aś* 'to pervade' (*cf.* Pāṇīnian *dhātupāṭha, aśū vyāptau* , 5th conj. *Ubhayapada*) with the suffix *an* and the *āgama t,* which is the same as above. The *sūtra* is: *sapy aśūbhyām tuṭ ca,* 4358 and the process is:

aś + an

= *aśt+an*

= *aśtan*

= *aṣṭan* (retroflexion of *śt*) = eight.

The root *aś* is from the 5th conj. and not from the 9th conj. (which means 'to eat') as the wording *asūbhyām* (with long *ū*) in the *sūtra suggests. The aś* from 9th conj. is designated by Pāṇini simply as *aśa* and *aśu.*

4.4.6. The words *ṣaṭ, nava* and *daśa* are nowhere derived though the *Auṇādikapadārṇava* refers to the *sūtras* in the commentary, which derive these three words; the *sūtras* are: *saheḥ ṣaṣ luk ca* , for the word *ṣaṭ* (i.e. ṣaṣ) and *nudaṁśor guṇaś ca,* for the words *nava* and *daśa*. The suffix seems to be *an*. And we have, for *ṣaṣ*,

sah+an

= *ṣaṣ+aṇ* (*sah* is substituted by *ṣaṣ*)

= *ṣaṣ+O* (*an* = o according to the above *sūtra*)

= *ṣaṣ* (= six).

For *nava,*

nu+an

= *no+an* (ū>o, *guṇa*)

= *nav+an* (ŏ>av, according to *eco'yavāyāvaḥ*)

= navan (= nine)

For *daśa,* the *sūtra* is: *nudaṁśor guṇaś ca.*

daṁś + *an*

= *daś+an* (the nasal *ṁ* = o, according to Pāṇini, 6.4.25: *daṁśasañjasvañjām śapi*)

= *daśan* (= ten)

As the remarks in the *Auṇādikapadārṇava* show, the above three *sūtras* are not available in any of the presently available recensions of the *uṇādisūtras; cf. T.R. Cintamani,* (*ibid.* p. 80):

asmin pāde "sapyaśūbhyām tuṭca" ityanantaram "nudaṁśor guṇaś ca" iti. "saheḥ ṣaṣ luk ca" iti ca sūtradyayam navan-daśan-ṣaṭ-iti śabdavyutpādakam ujjvaladattena vyākhyātam... evam anyair api tais tair vṛttikāraiḥ kānicit sūtrāny adhikāni vyākhyātāni sūtrakramabhedaś ca tatra bhūyān paridṛśyate pāṭhabhedāṅś ca bhūyāṁsa ity... alam bahunā.

The two number-words *dvi* and *catur* for 'two' and 'four' respectively are, however, even not referred to; nor again are they derived anywhere in any of the available recensions of the *uṇādisūtras; pañca* is only referred to. We thus have the linguistic derivation for 7 number-words, viz. *eka, tri, ṣaṭ, sapta, aṣṭa, nava* and *daśa*. Since the four words *sapta, aṣṭa, nava* and *daśa* are *n*-ending, the suffix assumed for them in the above grammatical process is–*an* i.e. a *n*-ending one; *pañca* is not derived at all. It is also to be noted that the suffix - *an* does not signify any meaning, unlike in Pāṇinī's grammar.

The above derivations of the *uṇādikāra* look very artificial and mechanical in the sense that there is no semantic equvivalence or relation between the meanings of the roots and that of the words derived from them. The meanings of the words thus derived would be: *tri* = 'the thing that swims or floats' (from *tṛ*); *sapta* = 'the thing which includes', and *aṣṭa* = 'that which pervades'. These meaning are not available for the words in the language. Even the suffixes, which in Pāṇīni's grammar guide us to the meanings of the derivatives, are not infused with any meaning at all.

Granting allowance for whatever deficiencies and differences from Pāṇini there are in *uṇādisūtrakārā's* etymologies, the problem which still remains is: why did the *Uṇādisūtrakāra* stop at giving out the etymologies of only 7 number-words? Why did he not attempt the etymologies of other words? He would have been at least taken as being consistent if he had etymologised the other remaining three words also.

4.5. *Underived words as base or radix for gradation*

We have seen above that grammatically the 20 words, viz. the first ten from *eka* to *daśa*, then all the nine multiples of *daśa* from *viṁśati* to *śata* and the number-word *sahasra* are underived words in the sense that no etymology or dissolution for them is given by any grammarian. All other number-words being compounds are or can be dissolved. This distinction between the two types of number-words from grammatical point of view can help us to understand why they are taken by the Vedic people as bases for gradations. The first gradation-mark is *daśa* and is underived. It, therefore, serves as the first base for gradation. And likewise all other underived words are automatically accepted as the bases for the next gradations. Except the number-word *daśa*, no other radix number-word is either derived or even referred to by any of the recensions of the *uṇādisūtras*.

5

The Methods of Countings

As we have seen before, the successive numbers are available by the addition of 1 to the immediately preceding ones. And the counting of the numbers also is done accordingly in that order, viz. *eka, dvi, tri .... saptadaśa, aṣṭādaśa* etc. But if we look at the Table No.1 which records all the numbers stated in the Vedic Saṁhitās, we find that for *navadaśa* (19) we have also the optional word-forms as *ekonaviṁśati* and *ekānnaviṁśati;* so also with other numbers like 29, which is expressed by *navaviṁśati* as well as by *ekānnatriṁśat.* For 39, we have only *ekānnacatvāriṁśat* and not *navatriṁśat. cf.* also forms for 49, 59, 69, 79, 89 and 99.

5.1. Forward and Backward Counting

We see from the above facts that the methods of counting numbers seem to be of two main types:

(a) Starting the counting from the lower level and climbing to higher level step by step and

(b) Starting from the higher level and descending to lower level so many steps as are required to arrive at the required number. Thus, the word *nava-daśa* (for 19) implies that the starting point of the count is *daśa* (10) and the ascending is done by nine (*nava)* steps to arrive at the required number 19. The words *ekonaviṁśati* or *ekānnaviṁśati* (or even *ūnaviṁśati* which is not available in the Vedic saṁhitās) indicate the starting point to be *viṁśati* (20) and the descending is done by one step to arrive at the required number. We may call the first method of counting as 'counting from the lower level' or 'lower-counting or under-counting' or even 'forward counting;' the second method then may be called 'the higher counting or the back-ward counting'. Thus the number *nava-daśa* (19) is expressed not only by 'forward counting' as *nava* (9) + *daśa* (10) but also by 'back-ward counting' as *viṁśati* (20) – (ūna) *eka* (one)'. To put it in other words, the units *eka, dvi* etc. are set in front of '*daśa'* upto *aṣṭādaśa;* but after that there is a sudden turn in the opposite direction and the units are placed before *viṁśati, trṁśat* etc. It must be remembered, however, that this type of backward counting is adopted—and that too optionally—in the case of the unit *nava* i.e. nine only. All other units are set only in front of or after *daśa.* In some languages, backward counting is adopted even in the case of units other than 'nine; *cf.* Lat. *duode-viginti* (=20-2). Later in the classical stage of Sanskrit, even the word *eka* is also dropped and the numbers are simply indicated by using the word *ūna* with the immediately following higher number; thus *ūna-triṁśat* is 29 and *ūna-catvāriṁśat* is 39. The use of *ūna* itself by convention means the subtraction of the number 1 and not any other number; the number 'one' (*eka*) need not be used. *K. Menninger* cites the examples of the Roman fractions in whose case the number 1 is taken as granted as being subtracted. Thus, in the fraction "11/12 de-unx (as - 12/12) from this ounce; 3/4 *dodrans<dc quadrans* "(as) from it 1/4." He tries to explain this tendency in counting in the following words:

"The next higher rank exerts its influence backward, dominating the numbers just preceding it, just as the full hour does the few minutes before it: "10 minutes to 6.00" rather than "5 hours 50." The rank levels, the old number-groupings, are numbers that early man could visualise........ 38 is just one of many numbers that are hard too grasp, but 40 is clear and palpable and so is 2; hence "2 from 40" can be understood whereas 38 cannot[8]." It is also to be noted in this connection that the numbers to be subtracted in backwarding counting from the next higher rank are never more than 3; that is to say, we do not get an example of, say, 35 obtained as 40-5 (in words as *pañca-ūna-catvāriṁśat*).

5.2. Counting by Multiplicaiton

Besides the above two methods, there is yet another method of denoting the numbers. We meet in the Vedic saṁhitās with expressions like *dviḥ pañca*, 'two times five' i.e. 'ten' (*cf.* RV 1.122.13: *dvir yat pañca bibhrato yanty annā*) or *triḥ sapta'*, 'three times seven' i.e. 'twenty-one' (=Cf. RV. 9.70.1: *trir asmai sapta dhenavo duduhre),* in which the number required is indicated as so many multiples of a certain base. Thus, in 1.122.13 above, 'ten' is represented as 'the second multiple' of the base five and in 9.70.1 "twenty-one" is represented as the third multiple of the base 'seven'. This method may be called as 'the method of counting by multiplication'. As we approach the Vedic *saṁhitās* later than the RV. we get frequent examples of counting by this method. But we find that this type of counting is not normally adhered to simply because of the fact that every number may or can not be expressed in terms of multiples of some other number. VS.17.2 quoted before is also an example of counting by multiplication of the previous number by 'ten' to hint thereby an infinite expansion of the series.

5.3. Counting by Indices or Ranks

VS.17.2. notes the following numbers: *eka* (1), *daśa* (10), *śata* (100) ..... upto *parārdha* (vide the discussion on this *ṛc* before) Each successive number that is noted is 'ten times' higher than the previous one. Besides indicating existence of counting by multiplication, the *rc* also implies the knowledge on the part of the Vedic people of counting by indices or ranks. Because, all the numbers noted there can be represented in modern notation in terms of the different ranks or indices of the base 'ten'. Actually this method of counting by multiplication or ranks was in ancient India used to test the knowledge of mathematics in general and of numbers in particular. The concept of indexing or ranking gave a wonderful tool in the hands of Vedic mathematicians, which helped them to expand the number—system *ad infinitum.*

Pāṇinian Description

We have said above that the numbers from 11 onwards are compound numbers mathematically since they contain two numbers giving out the sense of one concept. Though, therefore, the number 11, to cite an example, contains two digits symbolically, the concept which it signifies is a single concept of the number 11; it only implies that the number 11 succeeds the number 10. The same argument can be applied to all the compound numbers in mathematics.

This phenomenon of the number-symbols compounding with one another to express a single number is similar to the one found in languages like Sanskrit which abound in compound word-structures. Thus, to take an example, *rājan* is one word; puruṣa is another word; both convey two different meanings, viz. 'the king' and 'the servant'. Yet once they compound together in a word-structure like *rāja–puruṣa,* they convey a single concept, viz. 'the servant of the king[9]'. A more appropriate example would be of what is called a *bahuvrīhi* compound in Sanskrit, in which both the members in the compound totally lose their original meaning and give out a sense which is entirely different from the meanings of the members; *cf.* for example, a *bahuvrīhi* compound *pītāmbara* which is a compound of the two words *pīta* ('yellow') and *ambara* (= 'clothes, or waist-cloth'). The meaning conveyed by the entire compound word-structure '*pītāmbara*' (which is born out of the juxta-position of the two members), however, signifies a meaning '*viṣṇu*' (based on the dissolution, *pītam ambaram yasya saḥ,* 'one, whose waist-cloth is yellow') which is entirely different from the meanings of the two members. The mathematical phenomenon, therefore, of a compound number, though consisting of two number-symbols, signifying a single number-concept is very closely similar to the linguistic phenomenon of compound words in Sanskrit. Out of the four main types of compounds in Sanskrit. viz. *avyayībhāra, dvandva, tat-puruṣa* and *bahuvrīhi,* the possibility of the comparison with the mathematical phenomenon is provided by the compound *bahuvrīhi.*

But this is only a semantic or conceptual parallel between the mathematical and linguistic phenomenon of compound. Formally speaking, the formal parallel between the two phenomena is offered by the linguistic compound called *dvanda* in grammar when two words are juxta-posed and combined together by the meaning of 'and' i.e ca; *cf.* the Pāṇinian *sūtra, cārthe dvandvaḥ,* 2.2.29. The *dvandva*-compound of the two words, say, *rāma* and *lakṣmaṇa* is formed in the sense of '*ca*'; thus, *rāmaḥ ca lakṣmaṇaḥ ca* and the form of the compound-structure thus formed is *rāmalakṣmaṇa.* In the same way, as we have seen above, the compound numbers are formed by hypothising the meaning of '*ca*' i.e. 'and', *cf.* the Vedic passages quoted above in which the Vedas have stated the conjunctive particle '*ca*' which means 'and'. And we have *ekā ca daśa ca* = *ekādaśa* (=11); *dve ca viṁśatiś ca* = *dvāviṁśati* (=22); *dvau ca daśa ca* = *dvādaśa* (=12); etc. In the case of numbers above hundred, the function of *ca* i.e. 'and' is performed by the word '*uttara*', which literally means 'later or above', but signifies the general sense of 'and' in the matter of numbers. It must be noted, however, that the Vedas never use the compound of numbers above hundred; if they want to indicate the three-digit numbers (and also all numbers above three-digit rank), they split the numbers into one-digit or two-digit ranks and then use them; *cf.*, for example, the passages quoted above, viz. RV. 1.117.18 (*śatam ekaṁ ca* and not *ekottara-śatam*), RV.1.164.48 (*sākaṁ triśatā ṣaṣṭiḥ* and not *ṣaṭyuttaratriśatam*) etc. They thus seem to denote or count numbers by referring to the addition of different ranks and not by a compound, though all the numbers above daśa are compound numbers. The practice of denoting the numbers above 100 by a compound word-structure, and not by addition of different ranks, seems to be a post-Vedic phenomenon, developed in the classical stage.

Since the number-words above *daśa* are compound words, Pāṇinī, as a grammarian dealing with words, treats them as compounds, or rather as *samāhara dvandva* compounds and states the rules for the phonological/ morphelogical changes, if any, which take place during the process of compounding. Such phonological/ morphological changes occur only in the case of the first four number-words after ten, viz. the *ekādaśa* (=11), *dvādaśā* (=12), *trayo daśa* (=13) and *aṣṭādaśa* (=18). In the case of all other numbers no such changes are visible in the Veda. To explain, the compound of *eka* + *daśa* (for 11) *dvi* + *daśa* (for 12), *tri* + *daśa* (for 13) and *āṣṭa* + *daśa* (=18) should have formally been respectively *ekadaśa, dvidaśa, tridaśa and aṣṭadaśa.* But the Vedas use them as *ekādaśa* (with long *ā* of the final *a* of *eka*), *dvādaśa* (the final *i* of *dvi*>ā), *trayodaśa* (the number-word *tri*>*trayas*) and *aṣṭādaśa* (the final a of the word *aṣṭa*>ā) Pāṇini, therefore, as one having the greatest regard for the acceptability of the forms by the users or speakers of the language, had to explain these formations. According *to Brugmann* (*ibid.* III 25) the long *ā* is "instr. sing masc. (Ved-) nom. sing. femn.: the form thus chosen was suggested by *dvā-daśa.*"

5.3.1. *ekādaśa*

The final a of *eka* is substitued by ā (Paṇinian *āt*), the compounding of *eka* with *daśa* being according to the pāṇinian *sūtra, ān mahtaḥ samānādhikaraṇajātīyayoḥ,* 6.3.46; *cf. BD. ādiyogavibhāgād ātvam.* Thus *eka+daśa*=*ekādaśa* and not *ekadaśa.* The change of a to ā is also optionally explained by the *nipātana* of the word by Pāṇini in the *sūtra, prāg ekādaśabhyo' cchandasi,* 5.3.49; *cf.* B.D. *prāg ekādaśabhya iti nirdeśād vā.*

5.3.2. *dvādaśa* and *aṣṭādaśa*

The final sound is replaced by *ā* in the case of compounding of *dvi* and *aṣṭa* with *daśa* in all the number words except for the number-word before śatam. The *sūtra* is : *dvyaṣṭanaḥ saṁkhyāyām abahuvrīhyaśityoḥ,* 6.3.47. And we have *dvi+daśa* = *dvādaśa* and not *dvidaśa cf.* B.D. *dvyadhikāḥ daśeti; aṣṭa +daśa=aṣṭādaśa* and not *aṣṭadaśa*. The Pāṇinian *sūtra,* 6.3.47 is to be supplemented by the *vārttika, prāk śatād iti vaktavyam.* This phenomenon of the lengthening of *a into ā occurs only in the case of samāhāra dvandva* compounds and not in *bahuvrīhi* compounds and also not when *dvi* is compounded with *aśīti* (=80). The form with *aśīti* is *dvi-aśīti (=dyaśīti=82)* and not (*dvā-aśīti) dvāśīti*. The change to *ā* is optional for all the number-words from *catvāriśat* onwards. The sūtra *is; catvārimśat prabhṛti sarveṣām,* 6.3.49. And we have the optional forms *dvi-catvārimsat* and *dvācatvārimśat* (=42), *aṣṭa-catvārimśat* and *aṣṭā-catvārimśat* (=48), etc.[10]

5.3.3. *trayodaśa*

The word *tri* is replaced by the word *trayas* in the case of its *samāhāra-dvandva* compounds with number-words, excepting the number-word *aśīti* (=80). The sūtra *is,* tres trayaḥ, 6.3.48. We, therefore have *tri+daśa* = *trayas+daśa* = *trayodaśa* (=13). We cannot, however, have (*tri+aśīti*=) *trayo-aśīti*; we have only (*tri+aśīti*=) *tryaśīti.* The substitution of *trayas* for *tri* is, however, optional in the case of number-words from *catvārimśat* (=40) onwards; and we have the forms, *tricatvārimśat-trayaścatvārimsat* (=43), *tri-pañcāśat-trayaḥpañcāśat* (=53), *tri-ṣaṣṭi* or *trayaś-ṣaṣṭi* (=63), *tri-saptati* or *trayaḥsaptati* (=73) and *tri-navati* or *trayo-navati* (=93).

5.3.4. *ṣoḍaśa*

The word *ṣoḍaśa* signifies the number 16 and is derived from the compounding of *ṣaṭ* with *daśa*. Pāṇini has no *sūtra* in this respect. Yet the *vārttikakāra Kātyāyana* fills this gap and derives, by the *vārttika* ṣaṣa utvam *datṛdaśadhāsūttarapade ṣṭutvaṁ ca dhāsu ceti vācyam,* in the following way:

ṣaṣ + *daśa*

= *ṣa-u daśa* (final *ṣ* > *u*)

= *ṣo* + *daśa* (*a* + *u* > *o*)

= *ṣo* + *ḍaśa* (*d* > *ḍ*)

= *ṣoḍaśa,*

5.3.5. *ekānna-viṁśati and others*

It we go through the list of number-words given in Table No. 1, we find the following optional forms for the respective numbers. They are *ekānna–viṁśati,* and *nava-daśa* (for 19) and *nava-navati* and *ekānnaśatam* (for 99). The forms indicate the back-counting, starting from *viṁśati* (=20) and *śatam* (=100) with 'one' subtracted from them. The other forms which indicate back-counting are : *ekānna-catvāriṁśat* (40-1=39) (TS. 7.2.11.23), *ekānna-ṣaṣṭi* (60-1=59) (TS. 7.2.11.23), and *ekānna-aśīti* (80-1=79) (TS. 7.2.11.25). The corresponding forms for these numbers, based on forward-counting are: *nava-triṁśat* (9+30=39), *nava-pañcāśat* 59) and *nava-saptati* (9+70=79).

Pāṇini derives these structures in the following way linguistically/ grammatically. The *sūtra* is: *ekādis' caikasya cāduk,* 6.3.76. What Pāṇini does is that the number-words *viṁśati* (20), *triṁśat* (30), *catvāriṁśat* (40), *pañcāśat* (50), *ṣaṣṭi* (60), *saptati* (70), *aśīti* (80), *navati* (90) and *śatam* (100) are first compounded with the negative particle *na.*[11] The particle *nañ* i.e. *na* does not lose its initial *n,* but retains as it is. The negative compounds then would be *na-viṁśat,* na-triṁśat etc. When the word *eka* precedes this compound, the final *a* gets an augment viz. *ad* (Pāṇinian *aduk* in which u and *k*=o). And in the situation, *eka+navīṁśati,* the preceding or first member *eka* with the addition of *ad* assumes the form *ekād* (*eka+ad=ekād* by saṁdhi of *a+a* into *ā*); and the structure of the whole compound is *ekād-naviṁśati,* and by saṁdhi of final *d* of *ekād* and the initial *n* of *naviṁṣati* into *nn,* we have the form *ekānnaviṁśati*; so also *ekānnatriṁśat* etc. In the absence of the saṁdhi of *d* and *n,* the forms remain as *ekād-naviṁśati etc.* So we have two froms optionally—with or without saṁdhi of *d and* n, as *ekād-na-viṁśati* and *ekān-na-viṁśati etc.*

It should be noted, however, that the compounds *ekānnaviṁśati* etc. are *tatpuruṣa,* or to be exact, *tṛtīyā tatpuruṣa* compounds. They are to be dissolved as *ekena naviṁśati* etc. with the first member *eka* in the instr. case. The other corresponding words for these numbers 19 etc. are however, *samāhāra dvandva* compounds and not *tatpuruṣa*. Thus, *nava-daśa* (19) for example, is to be dissolved, and is actually explained in the Vedic texts, as *nava ca daśa ca*. The *tṛtīyā tatpuruṣa* compounds are brought about by the Pāṇinian *sūtra*, 2.1.30.

Besides the number-words *ekād-navimśati, ekānna-vimśati, nava-daśa* (for 19) etc. there is another word for the same numbers. And it is *ekonavimśati* for 19 etc. This word is a compound of three words viz. *eka, ūna* and *vimśati*, in which the word *ūna* means 'less than'; the whole compound means 'a number which is less than twenty by one.' This is also a back-counting. The compound is also a *tṛtīya tatpuruṣa* and not a *dvandva* and is to be dissolved as: *ekena ūnā* = *ekonā*; we compound this word *ekonā* with *vimśati* etc. as: *ekonā vimśatiḥ* = *ekonavimśati*. The first stage of the compound viz. *ekonā* is *tṛtīya tatpuruṣa* and the second stage is what is called as the *karmadhāraya compound* (*cf.* Pāṇini, 1.2.42). And we have the word *ekonavimśati* for 19 etc.

It also should be noted that the Vedas use only the first two options, viz. *nava-daśa* and *ekānna-vimśati* (19) and never the the third one, viz. *ekonavimśati*; the third word is found only in the Brachmanic and classical stage of Sanskrit.

Linguistically speaking *ekāt* in *ekānnavimśati* / *ekād-na-vimśati* seems to be abl. sing of the number-word *eka;* actually according to Pāṇinism *sūtra, sarvādīni sarvanāmāni,* 1.1.27, the number-word *eka* gets the *sarvanāma-saṁjñā,* and should get the termination *smāt* in the place of *āt* as the suffix for abl. sing. according to the Pāṇinian *sūtra, ṇasi-ṇyoḥ smāt-sminau,* 7.1.15, and the grammatically correct from for abl. sing. of *eka* should be *ekasmāt*. Though Pāṇini does not treat the form *ekāt* in the above word *ekān-na-vimśati,* as abl. sing., but treats it as one with the *āgama ad,* the form *ekāt* is actually found to be *ekasmāt* (as abl. sing.) in a passage from TS. which uses the form *ekasmāt-na-pañcāśa* (i.e. one less from i.e. than fifty) for the number 49th, which is an ordinal number for *ekasmāt-na-pañcāśat* i.e 49; *cf.* TS. 7.4.7 (*sa etam ekasmānnapañcāśam apaśyat).*

We have also the use of dat. sing-of *eka* in fem. for the same number 49 in TS. 7.4.7; cf. TS.: *tad yad ekasyai-na-pañcāsad atiriktāḥ...*

The use of cases other than the instr. sing. shows that the number-words coined by back-counting are not necessarily the *tṛtīyā tat-puruṣa* compounds but can also be taken as *caturthī tatpuruṣa* or *pañcamī tatpuruṣa* compounds. The use of this or that case seems to depend upn how one wants to signify the number—that is to say, whether by 'one less from' (abl.), or 'less by one for' (dat.) or 'a number less by one' (instr.).

6

The Ordinal Numbers

Besides the cardinal numbers noted above, the Vedic saṁhitās also use what are called as 'the ordinal numbers', which are used to define the things' position in the series. Thus, the cardinal number *'eka'* means 'one'; and the ordinal number *'prathama'* means the 'the first', which defines 'the first position of a thing' in the given series. We have the following ordinal numbers recorded in the RV. Saṁhitā. For the sake of brevity and avoiding unnecessary long list of the numbers, the ordinal numbers from only the RV. saṁhitā are noted down. They are as follows:

6.1. *prathama* (first), *dvitīya* (2nd), *tṛtīya* (3rd), *saptatha* (7th), *aṣṭama* (8th), *navama* (9th),, *daśama* (10th), *ekādaśa* (11th), *dvādasā* (12th), *pañcadaśa* (15th), *ṣaṭ-triṁśa* (36th) and *tri-pañcāśa* (53rd). There are many other ordinal numbers corresponding to their cardinal parts mentioned in different *saṁhitās* which are not listed here. There is a difference between the cardinal number-words and the ordinal number-words. While the cardinal number-words from *eka* to *nava* in Sanskrit are not derived words, but are underived or *siddha prātipadikas,* their ordinals are derived or *sādhita prātipadikas.* Except the *uṇādisūtrakāra,* no grammarian including Pāṇini takes the cardinal number-words as *sādhita* or derived. The ordinal number-words, however, are all derived ones from the point of view of all Sanskrit grammarians. Hence, it would be really interesting to know the derivational process by which the ordinal number-words are derived by Pāṇini. The Pāṇinian technical term for ordinality is *pūraṇa* (*cf.* the *sūtras,* 2.2.11, 5.3.48 etc.), though, it must be remembered that the idea of ordinality is nowhere defined by Pāṇini, nor again the word *pūraṇa* is anywhere defined—obviously because the idea is from outside the field of grammar.

6.2. prathama (first)

The ordinal number-word for the cardinal *eka* (=one) is *prathama.* The ordinal word *prathama* shows apparently no phonetic relation with its cardinal *eka.* Pāṇini has nowhere derived the ordinal word. Hence we have no other source to inquire about its derivation except in historical linguistics.

The derivaion of *prathama* from the point of view of historical linguistics goes back to the root **per* (from which, incidentally, the Sanskrit word *pūraṇa* for ordinality can be said to have been derived). This root * *per* gives out in Greek, Latin and other Western European languages words like *pro* or *protos* (meaning 'before', 'in front of' or 'first' in *Gk*), *pri* for *prae* (in old Latin, meaning 'before') as also *pri-or* ('one in front'), *primus* ('foremost, first') from which the French *premier* and Italian *primo* can be derived; so also English *premier* and *prime*; through change of *p>f,* we have the Germanic forms as the Gothic *frumists,* Anglo-Saxon *fornest,* Eng. *fore-st>first,* and German '*fūrst*' (= lord, prince). The same IE. base **per* on the other hand gives out in the Vedic Indo-Aryan words like *pūrva* (with the suffix - *va*) on the one hand and *pra-tha-ma* (with the suffix —*tama* (<* *tmmo,* IE) on the other. We have again the words *puras* 'before' and *purastāt* 'in front of'. The -*ta*- in the suffix - *tama* changes to -

tha-after the manner of the change to -*tha*- in *caturtha, saptatha* etc. Sk. *prathama* = Av. *fra-tema* = old Pers. *fra-tama.*[12]

Though Pāṇini does not derive the word, the *uṇādisūtrakāra* derives it from √*prath,* 'to spread' (*cf.* Pāṇinian *dhp. pratha prakhyāne,* 10 conj. UP); cf. the *uṇādisūtra, prather amac, 4952,* ; thus

prath + the suffix *amac*

= prath + ama (*c* = O according to 1.3.3 & 8)

= *prathama*

The suffix viz.—*ama* is applid to *car* to give the form *carama,* meaning "the last, the final"; thus *car* + *ama* = *carama; cf.* the *uṇādi-sūtra, cares' ca,* 4953.

6.3. ***dvitīya* (second)**

The Pāṇinian *sūtra,* dves *tīyaḥ,* 5.2.54, lays down the suffix *tīya* for the word *dvi* and we have,

dvi + *tīya*

= *dvitīya.*

The sense of the suffix is obviously *pūraṇa.* Linguistically, Sk. *dvitīya* = Av. *bitya* = Gāthic *dabitya* = O pers. *dūvītīya.* According to K. Brugmann, the Gk. *Deuteros* (= the one removed from something i.e. second) is to be connected with Sk. *dūra* and not with *dvi.*[13]

6.4. ***tṛtīya* (third)**

Pāṇini derives it from the base *tri* with the suffix *tīya;* in this process, the *r* of *tri* undergoes *samprasāraṇa.* Thus, *tri* + *tīya* = *tṛ* + *tīya* = *tṛtīya.* The Pāṇinian *sūtra* is: *treḥ saṁprasāraṇam ca,* 5.2.55.

It is to be read with 5.2.54 quoted above. Linguistically. Sk. *tṛtīya* = Gr. *tertos* = Lat. *tertius* = Pruss. *tirtis* = Umbrian *tertim;* the *ter* = **tṛ.* The Indo-Germanic languages thus have *to* for the sk. *ti* in -*tīya.* The Av. and O Pers. show a difference. Av. = *thritya* and O

Pers. = 'sritya, *the Sk. tīya* losing its middle *i* and changing to *tya.*[14] Gothic has *thridja* and OHG. *dritto.*

6.5. caturtha (fourth)

Pāṇini in the *sūtra, ṣaṭ-katipayacaturām thuk,* 5.2.51, states the *āgama thuk* i.e. *th* (*u* and *k* both = O) before applying the *pūraṇa*-suffix *ḍaṭ* i.e. a (both *ḍ* and *ṭ* = O) to the word *catur* 'four'. The suffix *ḍaṭ* is given in the *sūtra, tasya pūraṇe ḍaṭ,* 5.2.48. The process, therefore, to arrive at the ordinal number-word *caturtha* is:

catur + a (i.e. *ḍaṭ*)

= *catur* + *th* + a

= *caturtha.*

Since the *āgama thuk* is a *kit* i.e. with *k* as the *it*-sound, (and hence zero), it is to be applied at the end of the word *catur* according to the *sūtra, ādyantau ṭakitau,* 1.1.46.

The word *catur* also gives out two more ordinal words, viz. *turīya* and *turya.* Pāṇini has not given any process to arrive at these two derivatives, which short-coming is covered up by Kātyāyana, the *vārttikakāra.* The *vārttika caturas' cha-yatau ādyakṣaralopas' ca* is on the Pāṇinian *sūtra* quoted above. The *vārttika* applies the two suffixes *cha* (which is substituated by *īya* according to the *sūtra,* 7.1.2) and *yat* in the sense of *pūraṇa* and while applying these two suffixes, the base *catur* loses its initial sounds *c* and *a.* Thus the process is:

(1) *catur* + *cha*
= *catur* + *īya* (*ch* > īya; *cf.* 7.1.2)
= *tur* + *īya* (*ca* = O)
= *turīya*, and

(2) *catur* + *yat*
= *catur* + *ya* (*t* =O)
= *tur* + *ya* (*ca* = O)
= *turya.*

We have thus three ordinal forms, viz. *caturtha, turīya* and *turya* for the cardinal number-word *catvar* > *catur.* It is to be noted that while applying the suffix *cha* and *yat* given by the vārttika, *the Pāṇinian suffix* ḍaṭ i.e. a is not to be applied.

Linguistically, though all the Sk. forms, viz. *caturtha, turīya* and *turya* are equal to Av. *tūirya* only the Sk. words *turīya* or *turya* seem to be the real parallels for the Av. *tūirya*. Gk. has *tetartos* and Latin shows *quartus.*[15] The Sk. *āgama tha* (Pāṇinian *tha* or *thaṭ* for which see below) is continued in English even to-day; and we have in English the ordinal words as 'fourth, fifth, sixth' etc.

6.6. *pañcama* (fifth), *saptama* (seventh), *aṣṭama* (eighth), *navama* (ninth) and *daśama* (tenth):

The number-words *pañca, sapta, aṣṭa, nava* and *daśa* are all designated in Sanskrit grammar by Pāṇini as ending in *-n*; and we have their basic *prātipadika*—form as *pañcan, saptan, aṣṭan* (*cf.* the *sūtra, aṣṭanaḥ ā vibhaktau,* 7.2.84, which states the gen. sing. of *aṣṭan as* aṣṭanaḥ and not *aṣṭasya*) *navan* and *daśan.* Not only this, but even the words like *ekṣdaśa, dvādaśa* etc. which have *daśa* as the second member are also taken to be *n*–ending. The Pāṇinian *sūtra, nāntād asaṁkhyāder maṭ,* 5.2.49 states the suffix *maṭ* i.e. *ma* (with *ṭ* = O) in the sense of *pūraṇa* i.e. ordinality for the number-words. The suffix *ma* (i.e. *maṭ*) replaces the general suffix a (i.e. *ḍaṭ*) and we get,

pañcan + *ḍat*

= *pañca* + *a*

= *pañca* + *ma* (a replaced by *ma*)

= *pañcama.*

If, however the *n*–ending number-words occur as the first member of a compound, the suffix *ma* is not to be applied; *cf.* the word *pañca-daśa* (= fifteen) etc. in which the word *pañca* occurs as the first member. The word *pañca,* therefore, is not eligible to get the suffix –*ma* and to give out the ordinal number - form as *pañcama-daśa* . or even *pañca-daśama.*

The same is the case with the other *n*–ending number-word and we have their ordinals as *saptama, aṣṭama, navama* and *daśama,* if they occur independently and not in any compound word-structure as the first member.

Of all the ordinal number-forms, the form for *sapta* exhibits an optional variety, viz. *saptatha;* the other word which shows a *tha*-ending ordinal number - structure is *ṣaṣṭha* (sixth) from *ṣaṣ* (six). The optional form *saptatha* is available only in the Vedic language and theretoo only in the RV. (*cf.* 1.164.15; 10.99.2 and as fem. *saptathī,* 7.36.6). Pāṇini notes this phenomenon and frames a general rule which is practically applicable only in the case of the words *ṣaṣ* and *sapta,* but which theoretically can also be applied in the case of all other *n*-ending number-words. The rule is : *thaṭ ca chandasi,* 5.2.50. The rule, which succeeds the rule 5.2.49 (laying down the substitution of *ma* for a), states that the suffix -*tha* (i.e. *thaṭ,* with *ṭ* = O) along with the suffiix -*ma* also takes the place of a (i.e. *ḍaṭ*) in the Veda. And we have,

sapta + thaṭ

= *sapta + tha*

= *saptatha*

So also with other number-words and we have the optional ordinals as = *pañcama/pañcatha, saptama/saptatha, aṣṭama/aṣṭatha, navama/navatha* and *daśama/daśatha.*

K. Brugmann[16] notes this phenomenon. Av. has both the suffixes for *pañca*; *cf.* SK. *pañcama* = Av. *paṇtaṇhem* and Sk. *pañcatha* = Av. *puxtha. cf.* Gothic *fimfta;* OHG. *fimfto.* All these can be traced back to Indo-Germanic* **pṁq-to* or * *peṁq-to.*

Indo-Germanic **septmo* or *septṃmo* gives out SK. *saptama,* Av. *haptama,* Pruss *septma.*

Indo-Germanic **oktoṃ-o* = SK. *aṣṭama* = Av. *aštema.*

Indo-Germanic * *neuṇn-o* or * *neuṇto* = Sk. *navama,* = Av. *naoma* = O Pers. *navama,*

Indo-Germanic **dekṁto* or **dekm̥ mo* = Sk. *daśama* = Av. *dasema,* = Lat. *decimus.*

The *tha,* as the optional suffix for the ordinal follows the other branch of Anglo-Saxon and is visible even today in English as in 'fifth, sixth' etc.

6.7. *ṣaṣṭha* (sixth)

The Pāṇinian *sūtras* 5.2.48 and 5.2.51 lay down the suffixes a (i.e. *ḍaṭ*) and *th* (i.e. *thuk*) respectively for the word *ṣaṣ* to give out its ordinal. And we have, *ṣaṣ* + a = *ṣaṣ+th+a* = *ṣaṣṭha.*

The Indo-Germanic *suektos = SK. *ṣaṣṭha*= Av. *Xśtva* = Lat. *sextus* = OHG *sehsto/sehto* = *Goth.* saihsta.[17]

6.8. Ordinals for number-words above *daśa*

The Pāṇinian general rule for forming the ordinal number-words above *daśa* is : *tasya pūraṇe ḍaṭ,* 5.2.48, which states the suffix *ḍaṭ* i.e. a to all the number-words. Thus from the number-word *ekādaśa,* we have,

ekādaśa + *ḍaṭ*

= *ekādaśa* + a (*ḍ* and *ṭ* = O)

= *ekādaśa* + a (final *a* = O according to 6.4.143)

= = *ekādaśa,* 'eleventh'.

Since the word *ekādaśa* thus formed ends in -*a,* it is declined like any other *a*-ending nominal bases and not like the *n*-ending nominal bases. Thus, the nom. plu. of *ekādaśan,* the *n*-ending cardinal, will be *ekādaśa* (*cf.* RV. 1.139.11) while that of *ekādaśa,* the *a*-ending ordinal, will be *ekadaśāsaḥ* (RV. 8.57.2; 9.92.4 etc.). The suffix *ḍaṭ* being for general application, the succeeding number-words from *ekādaśa* onwards will be formed in the same way; so we have *dvādaśa* (= twelveth), *trayodaśa* (= thirteenth), *viṁśa* (=twentieth), *ekaviṁśa* (=twenty-first), *triṁśa* (=thirtieth) etc.

The cardinals *vimśati, triṁśat, catvāriṁsat, pañcāśat, ṣaṣṭi, saptati, aśīti, navati* and *śatam,* however, exhibit their ordinals with the suffix –*tama* also. Pāṇini, therefore, in a special rule, 5.2.56 (*viṁśatyādibhyas tamaḍ anyatarasyām*), prescribes the suffix *tamaṭ* i.e. *tama* (*ṭ* = O) for all the words noted above. And we have the optional ordinals for these words as : *viṁśatitama,* (20th), *triṁśattama* (30th), *catvāriṁṣattama* (40th), *pañcāśatama* (50th), *ṣaṣṭitama* (60th), *saptatitama* (70th), *aśītitama* (80th), *navatitama* (90th) and *śatatama* (100th).

The rule also implies that not only the words noted in the *sūtra* are only eligible to get optionally the suffix *tama,* but also all other words ending in these words also get optionally the suffix. Thus, to cite an example, the word *eka-catvāriṁsat* (41) ending in *catvāriṁśat* will show two forms for its ordinal as *eka-catvārṁśa* (with the suffix *ḍaṭ* i.e. a) and *eka-catvāriṁśat-tama* (with the suffix -*tama* = 41st).

Though the suffix *ḍaṭ* i.e. *a* is of general application, it is not seen to have been applied to *i*-ending cardinals from *ṣaṣṭi* onwards, viz. *ṣaṣṭi, saptati, aśīti,* and *navati.* These words get only the suffix -*tama.* The *i* -ending word *viṁśati,* however, exhibits its ordinal with both the suffixes, as *viṁśa* (TS. 7.3.9.2) as well as *viṁśatitama.* So also *triṁśat, catvāriṁśat* and *pañcāśat.*

Linguistically also, all these numbers from *ekādaśa* to *navadaśa,* which end in -*daśa* "have both -*daśa-s* and -*daśama-s;* cp. Lat. -*decimu-s,* 11th Skr. *ekādaśa-s,* Avesta *aevan-dasa, aeva-dasa; aeva-dasa* may be like *dva-dasa* = Skr. dvā-dasa." SK. *dvā-daśa* = Av. *dvadasa,* = Gk. *dodekatos,* or *duo-dekatos* = Lat *duo-decim.*

SK. *trayo-daśa* = Av. *thridasa.* SK. *caturdaśa* = Av. *cathrudasa.* SK. *pañca-daśa* = Av. *panca-dasa* and *panca-dasya;* SK. *ṣoḍaśa* = Av. *Xśvaś-dasa.* SK. *navadaśa* = Av. *navadasa.* The Lat. parallels are : *ūndecimus* (11th), *duodecimus* (12th), *tertiusdecimus* (13th), *quartusdecimus* (14th) etc. Rarely we find *decimus tertius* (for

13th) and *octāvos decimus* (for 18th) in the place of their regulars, viz. *tertius decimus* and *duo-de-vicesimus.* In Avesta, only *thrisata* for Sk. *triṁśa* (30th) is found; others are not available; for details, *cf. K. Brugmann, ibid.* III p. 35 ff.

7

Numbers without Number-words

Besides the actual number-words which, in the form of some symbols, are and can be used in mathematics, not only Sanskrit, but every language contains certain words which contain—and donot convey directly—the number-concept; they convey a number-sense, but are not explicity number-words. To cite an example from English, the word 'many' is not a number-word signifying any number; yet it has a numerical signification. The same phenomenon can be observed even in the Vedic language. The following are some of the examples of words which are not numbers but which implicily convey a number-sense.

7.1. *anya*

The word *anya* primarily means 'the other', which meaning implies the presence of something which is 'the one or the first.' It, therefore, has come to signify a sense of 'the second'; *cf.* RV. 1.164.20: *tayor anyaḥ pippalam svādv atti, anaśnann anyo abhicākaśīti,* in which *anya* is contrasted with 'one' and hence means 'the other than the one'.

Perhaps the derivation of the word may go back to the IE base **oi-no-s* which means 'one' in Lat. '*oi-nos, oeno-s, ūnus'*; old Irish '*oe-n*'; German '*ein*' and Eng. 'one'. K. Brugmann connects it with SK. *ena* 'he'.[18]

Along with *anya* go the words *anyatara* (comparative) and anyatama (superlative) also.

7.2. Ubha or ubhaya

The word means 'two things together in pair' i.e. 'both'. While the word *ubha* is declined in all the three genders only in the dual number, the word *ubhaya* is declined in all the three numbers and genders. Examine, for example, the dual number of *ubha* in passages like: RV. 1.22.2: *ubhā devā divispṛśā,* (masc. but only du.) RV. 1.33.9: *ubhe dyāvāpṛthivī* (fem - du) and *janmanī ubhe* (neu. du.), RV. 1.141.11. We have the following forms of the word in the RV: *ubhau* (nom-acc. masc. du.), *ubhe* (nom. acc. fem. neut. du), *ubhābhyām* (instr., dat., abl. du., masc, fem. neut.) and *ubhayoḥ* (gen. loc. du., masc. fem. neut.).

The word *ubhaya* is found in all numbers and genders. The occurrences are *ubhayam* (masc. neut sing.), *ubayasya* (masc. gen. sing.), *ubhayā* (masc. du. nom. acc.), *ubhayāḥ* (masc. fem. plu.), *ubhayān* (acc. masc. plu.), *ubhayāni* (nom. acc. plu., neut.), *ubhayāya* (masc. dat. sing.), *ubhayāsaḥ* (masc. fem. nom. plu.) *ubhaye* (neut. fem. nom. du.), *ubhayebhiḥ* (masc. instr. plu.), *ubhayeṣām* (masc. gen. plu.), *ubhayeṣu* (masc. loc. plu.) and *ubhayoḥ* (masc. fem. neut. gen. loc. du.). *cf.* BD. on the *sūtra,* 7.1.52: *ubhaśabdaḥ dvitvaviśiṣṭa-vācakaḥ; ata eva nityam dvivacanāntaḥ.* As regards the dual of *ubhaya,* there are two opinions. According to Kaiyyaṭa, the word *ubhaya* is not to be necessarily used in dual; according to Haradatta, dual is necessary; *cf.* BD.: *ubhaya-śabdasya dvivacanam nāstīti Kaiyyaṭaḥ; astīti Haradattaḥ.* The word is derived by Pāṇini from *ubha* with the suffix *ayac* i.e. *aya* (< *tayap*); *cf.* the *sūtra, ubhād udātto nityam,* 5.2.44. It means 'a group of two.'

7.3. etāvat

It means 'this much'; but in numerical context, it signifies 'so many'; *cf.* TS. 3.5.2: *aṣṭau vasavaḥ; ekādaśa rudrāḥ; dvādaśādityāḥ etāvanto vai devāḥ;* Pāṇini derives the word from *etat* with the suffix *-vat* i.e. *vatup*. Thus, *etat + vatup = etat + vat = eta + vat = etāvat; cf.* the *sūtra, yat-tad-etebhyaḥ parimāṇe vatup,* 5.2.39.

7.4. puru

The word means 'many' and is used in the same sense in all the occurrences in the *saṁhitās*. The context in which it occurs are many, which can be classified into two main types - numerical and non-numerical. In numerical context, it signifies the numerical content; *cf.* RV. 1.62.10: *purū sahasrā janayo na patnīḥ;* RV. 1.81.7: *saṁ gṛbhāya purū śatā,* etc.

7.5. kiyat

It means 'how many' or 'how much'. In numerical context, it has a numerical significance; *cf.* RV. 10.27.12: *kiyatī yoṣā ...vadhūr bhavati,* Pāṇini derives the words from *kim* with the suffix *gha* (> *iya,* according to 7.1.2) which replaces the *va* of the suffix *vat* i.e. *vatup* according to the *sūtra, kimidaṁ-bhyām vo ghaḥ,* 5.2.40. Thus, *kim+vat = kim+ghat = kim + iyat = k + iyat = kiyat.*

7.6. kati

It means, in question, 'how many', which implies counting and calculating. This sense is visible in the Ṛgvedic passage 10.88.18: *katy agnayaḥ kati sūryāsaḥ, katy uṣāsaḥ katy u svid āpaḥ; cf.* RV. 9.72.1; 10.86.20. The formation *kati—dhā* ' in how many ways' occurs in the RV. 1.31.2; 10.90.11. The answers to the questions containing the question-word *kati* are always given or expected in numbers. Pāṇini derives *kati* from *kim* with the suffix *-ati* (Pāṇini *ḍati*); thus *kim + ḍati = kim + ati = k + ati = kati, cf.* the *sūtra,* 5.2.41: *kimaḥ saṁkhyā-parimāṇe ḍati ca.* The word *saṁkhā-parimāṇa* is notable.

7.7. bahu

The word means primarily 'many, plenty of' etc. *cf.* RV. 2.18.3: *bahavo hi viprāḥ.* The word implies a plural number—an indefinite plural. *Pāṇini* also uses the word *bahu* in the *sūtra, bahuṣu bahuvacanam,* 1.4.21 to signify the sense of 'many' i.e.

'three or more than three'—an idefinite plural. So also are the words *bhūyas* and *bhūyiṣṭha* which signify the comparative and supelative degrees of *bahu*.

Besides the above words which are not themselves number-words but which signify number-sense, there are others like *ādi, ādya* (both meaning 'the initial'), *agra* (= in the fore or first position'), *anta* and *antima* (= 'the last, the final') etc. which imply numerical context. These words, however, donot occur in the numerical context in any of the Vedic *saṁhitās* taken here for study. Hence they are dropped from discussion. They occur frequently in the numerical sense in the classical literature. But one thing is certain. Though these words convey a numerical meaning, they do not seem primarily from the numerical context; they seem to have been taken over from non-numerical context. The words *yāvat, tāvat* and *iyat* also seem to convey the numerical signification, but it is very faintly deciphered in the single occurrence. For *yāvat—tāvat* (= 'as many .... so many', *cf.* TS. 6.1.47: *yāvataḥ eva paśūn abhi dīkṣeta tāvantaḥ asya paśavaḥ syuḥ;* for *iyat, cf.* VS. 10.25 (which is repeated in TS, MS; kāṇva and KS): *iyad asi, āyur asi* where Uvaṭa remarks: *iyad iti parimāṇavacano'yam śabdaḥ* and paraphrases as *śatamānam asi:* Mahīdhara says: *etāvatparimāṇam śataraktikāparimitam asi .... tasmād āyuḥ śatābdaparimitam mayi dhehi; yatas tvam śatamānam asi tataḥ śatābdaparimitam āyur mayi ropaya.*

7.8. carama

Derived from √*car* with the suffix -*ama,* the ordinal formation *carama* signifies 'the last, the final'. Though not a number-word itself, it does signify a numerical meaning in numerical contexts; *cf.* RV. 7.59.3, 8.20.14, 61.15 etc. It is thus semantically equivalent of *antima* from *anta,* 'the end'.

7.9. asaṁkhyāta

Derived from *a+sam+* √*khyā*+the past pass participial suffix; the word means 'that which are not counted' i.e. innumerable. The word occurs in VS. 16.54: *asaṁkhyātā sahasrāṇi ye rudrā adhi bhūmyām.*

8

Numbers as Adjectives

The number-words, being primarily words i.e. nominal bases; as such they are all inflected in Sanskrit which is highly inflected. As such, just as the termination or *pratyaya* (referring to the closing morphemes) is a compulsoy category in Sanskrit; so also, the gender and number of a word are also compulsory, so far as the inflection of the word is concerned.[19] A declined Sanskrit nominal base therefore, indicates the number, the *pratyaya* (meaning the case) and the gender simultaneously. Thus, when the declined nominal base *indraḥ* or *marutvān* is uttered, it is immediately known that the nominal base in question is singular in number, a masculine and is used in first i.e. nominative case. What is still more interesting to note is that out of these three categories again, it is the category of *prataya* i.e. termination, referring to the closing morphemes[20], which helps us to know the gender and number of the nominal base. Thus, in *indraḥ* or *indrasya,* the termination *-aḥ* (i.e *su* in Pāṇini's grammar; *cf.* the *sūtra* 4.1.2) or *sya* indicates that the form is masc. nom. sing. or masc. gen. sing. respectively.

The number-words, being nominal bases, are also, therefore declined in Sanskrit. We will now, therefore, discuss the number, gender and the cases of the number-words as found in the Vedas.

8.1. Number (*vacana*) of the number-words

There are three number (*vacana*) in Sanskrit, viz. singular, dual and plural, defined by Pāṇini in the *sūtras, dvyekayor dvivacanaikavacane,* 1.4.22. and *bahuṣu bahuvacanam,* 1.4.21 (for plural or more specifically, indefinite plural). The Pāṇinian terms for singular, dual and plural are respectively *ekavacana, dvivacana* and *bahuracana.* Since Sanskrit is an inflected language, the *vacana* is indicated by or resides in the termination applied to the nominal base. The different terminations of the seven or eight *vibhaktis* (cases) are listed and stated by Pāṇini in the *sūtra,* 4.1.2.

8.1.1. *eka*

Since the number-word *eka* signifies singularity, one would expect it to be declined in only one way i.e. with the suffix of only the singular number in all the seven/eight *vibhaktis,* as is the case with the number-word for 'one' in IE languages like Greek and Latin.[21] But it is not so in Sanskrit. Even the number-word '*eka*' denoting 'singularity' is declined in plural number in as old times as those of the Vedas themselves; *cf.* RV. 8.29.10: *arcanta eke mahi sāma manvata;* 10.114.10: *bhūmyā antam pary eke caranti;* and 10.154.1: *soma ekebhyaḥ pavate, ghṛtam eka upāsate,* in which the form *eke* is nom. plu. and *ekebhyaḥ* is dat. abl. plu. It should, however, at the same time be remembered that in the above as well as all other contexts in which the word *eka* is declined in the plu, the word undergoes a little semantic change from simply 'one', to 'some,' used elliptically, some such word as 'some people, beasts, things' etc., to be supplied.

8.1.2. *dvi*

The number-word '*dvi*', however, as one would obviously expect, takes always the dual number and is declined as such. So we have only the three forms for the seven *vibhaktis* of the word as *dvau* (nom, acc.), *dvābhyām* (instr., dat., abl.) and *dvayoḥ* (gen., loc.). This is in line with its declension in the other IE. languages like Greek and Latin. Only the word *eka* strikes a difference in Sanskrit. Thus we have *dvau* (nom. acc.), RV. 1.131.3 etc; *dvābhyām* (instr., dat., abl.), RV. 2.18.4 etc and *dvyoḥ* (gen., loc.) RV. 1.83.3 etc.

8.1.3. *tri and other higher numbers*

All the other successive number words are always declined in plural. This plural is indefinite plural; it means anything 3 or beyond 3. We thus have *trayaḥ, catvāraḥ,* etc. They donot get the singular suffix. So we have, to cite the example of masc. *tri, trayaḥ* (nom.), RV. 1.34.2 etc, *trīn* (acc.) RV. 1.126.5 etc; *tribhiḥ* (instr.), RV. 1.34.11 etc.) *tribhyaḥ* (dat. abl.), RV. 10.185.1 etc. and *triṣu* (loc.) RV. 1.15.4 etc. The same is the case with all other successive numbers.

We thus see that so far as the number of the first three number-words, viz. *eka, dvi,* and *tri* is concerned, the word *eka* is declined in both, i.e. singular and plural number, and *never* in dual; the word *dvi* is never used in singular and plural and the word *tri,* always in plural, is never used in singular and dual. All other number-words following *tri* follow the same rule as the word *tri does,* except the number-words *śatam, sahasram* and all those in their multiples of hundred, which are declined in all the three numbers viz. singular, dual and plural.[22]

8.2. The gender of the number-words:

If we examine the texts of the Vedas, we find that the number-words *eka, dvi, tri* and *catur* exhibit declension in all the three genders. Thus, in RV. 1.7.9 (*ya ekaś carṣaṇīnām*) *eka* is used in masc. qualifying *indraḥ;* in R.V. 3.7.2 (*pary ekā carati vartanim gauḥ*), as the form *ekā,* qualifying *gauḥ,* 'indicates, it is used in feminine; in RV. 1.93.4. (*avindatam jyotir ekam bahubhyaḥ*), *eka,* qualifying the neut. *jyotiḥ,* is used in neuter gender.

So also the number-word ḍvi, which is used in all the three genders, viz masc. *dvau* (*cf.* RV. 1.35.6 etc.), fem. *dve* (*cf.* RV. 1.95.1) and neut. *dve* (*cf.* RV. 1.155.5 etc.). The number-words *tri* and *catur* are also used in all the three genders; *cf.* for masc. *tri,* RV. 1.34.2 etc.; for fem. RV. 1.13.9 etc.; for neut. RV. 1.22.18 etc. For masc. *catur, cf.* RV. 1.122.15 etc.; for fem., *cf.* RV. 1.62.6 etc. for neut. *cf.* RV. 1.164.45 etc. It is not only the context and the declensional terminations, but even the *prātipadika-* forms themselves that lead us to know the gender of th above words. The word *eka,* an a-ending masc, changes to *ekā,* ā-ending fem. The Pāṇinian *sūtra* which changes *eka* to *ekā* is *ajādyataṣṭāp,* 4.1.4. The word *dvi* does not undergo any change in its *prātipadika* - form in fem. (the *prātipadika*is defined by Pāṇini in 1.2.45). It is only the declension and the context which gives us a clue to know the gender of *dvi.* The word *tri,* masc. changes to *tisṛ* in fem; the Pāṇinian *sūtra* is *tri-caturoḥ striyām tisṛcatasṛ,* 7.2.99. It is this same *sūtra* which also changes *catur,* masc., to *catasṛ* in fem.

The number-words from *pañcan* to *daśan* do not change at all either in their *prātipadika* stage or even in the declensional stage so far as the gender is concerned. Thus, for example, we can have *pañca* in the same *prātipadika* - form and declined structure as *pañca* in masc. (as in *pañca janāḥ,* RV. 1.89.10), fem. (as in *imāḥ pañca pradiśaḥ,* RV. 9,86.29) and also neut. (as in *pañca padāni,* RV. 10.13.3). The same is the case with all other number-words upto *daśa.* In the case of the words *ekādaśa* onwards also, their gender is to be conjectured on the basis of the gender of the substantive they qualify. Though they are declined, they are in a sense genderless. Thus, in *ekādaśa devāḥ* (RV. 1.139.11), the words *ekādaśa* is to taken as masc.; in *ekādaśa* brāhmaṇācchaṁsyaḥ (KS. 21.12), it is fem and in *ekādaśa akṣarāṇi,* (MS. 1.11.10), it is neuter.; *cf.* also the word *śata* which has no gender by itself, the gender depending on its substantive; thus in *śataṁ kumbhān* (RV. 1.116.7), it is masc; in *śatam śaradaḥ* (RV. 2.27.10) it is fem. and in *śataṁ rādhaḥ* (RV. 4.31.9) it is neut. But when the word *śatam* is declined, it is always declined in neut., plu. irrespective of the gender of its substantive; *cf; śatā ..puraḥ,* fem. (RV. 1.53.8); *śatā vasu* neut)RV. 1.81.7); *harayaḥ śatā,* masc. (RV. 6.47.18); or even if there is no substantive, as in *śatā enam anonavuḥ,* RV. 1.80.9. The same is the case with the word *sahasram.* However, when *śatam* itself acts as a substantive, being qualified by number - words, the number is either sing. or plu.; but the gender is neut; *cf. trīṇi śatā* (RV. 5.29.8 etc.) or *daśa śatā* (RV. 5.62.1 etc); so also with the word *sahasra.*

Some of the number-words after *ekādaśa,* however, have fixed genders in absolute sense, i.e. in their *prātipadika-form.* This is the case with number-words ending in *-ti,* such as *viṁśati, ṣaṣṭi, saptati, aśīti* and *navati.* They are always in feminine gender, when they are to be qualified by an adjective or pronoun; thus we have *dve viṁśatiḥ,* TS. 5.3.3, and not *dvau viṁśatiḥ.* These words in *-ti* are always declined in singular number as in *dve viṁśatiḥ* quoted above. Actually, if viṁśati is to be qualified by *dve,* it should be *dve viṁśatī;* but it is not so. The same is the case with other number-words ending in *-ti.* It may be due to the fact that almost all Sanskrit *prātipadikas* ending in *-ti* are feminine in gender. This fact is noted in the *liṅgānuśāsana-sūtra, viṁśatyādir ā navateḥ,* No. 13. The *sūtra,* however, includes even the words *triṁśat, catvāriṁśat* and *pañcāśat* also, although they do not end in *-ti.* The Vedic evidence in the case of the feminine genders of the non-*i*-ending three words, however, does not substantiate the observation of the *sūtra.*

What is more interesting to note is that unlike in the case of the first four number-words, viz. *eka, dvi, tri* and *catur* which follow the substantive in number, gender and case other number-words do not necessarily take over the case and gender of their substantive; *cf.* for example, the following situations in which the number-words *pañca* and *sapta* do not follow the case of their substantive; *pañca* kṛṣṭīnām *(for* pañcānām kṛṣṭīnām, *RV. 1.7.9),* sapta dhāmabhiḥ *(for* saptabhiḥ dhāmabhiḥ, RV. 1.22.16) etc. The declensional terminations are in these cases, to put it in Pāṇinian terminology, zeroed. Pāṇini in the *sūtra,* 7.1.39, zeroes these terminations. Yet, this *sūatra* is applicable in the case of the zero of the terminations of only the *substantives* and not the adjectives, as in *parame vyoman,* in the place of *parame vyomani;* and that too at *pāda-end; cf.* BD's commentary and examples thereon. The rule 7.1.39, therefore, seems inapplicable in our present case. Even Kātyāyana does not add any *vārttika* to that effect to impove on Pāṇini. Does it mean that the phenomenon went unnoticed by both Pāṇini and Kātyāyana? The problem requires to be studied independently. Yet the fact remains that the number-words from *pañca* onwards donot show the declensional terminations applied to them in all the cases: Does it mean that the number-words from *pañca onwards* as adjectives were never declined?

8.3. Number-words as adjectives:

But we do get examples in the Vedic texts themselves in which the number-words as adjectives were declined in the same case as their substantives; *cf;* for example, *janeṣu pañcasu,* RV. 3.37.9, 9.65.23; *saptabhiḥ putraiḥ,* RV. 10.72.9;; *aṣṭau kakubhaḥ,* RV. 1.35.8 etc. This only shows that the phenomenon of application of declensional terminations to the number-words as adjectives occurring together with substantives was voluntary and the rule to that effect was optional.

This is about the cardinal number-words. The ordinal number-words like *prathama, dvitīya ..... pañcama .... daśama* etc. however, follow without any exceptions the number, gender and the case of their substantives. This shows that both the two-types of number-words are full-fledged adjectives.[23] They, therefore, followed the rule laid down for the adjectives viz.

yal liṇgam yad vacanam,
yā ca vibhaktir viśeṣyasya
tal liṇgam tad vacanam,
saiva vibḥaktir viśeṣaṇasyāpi

Yet, they are not adjective in the sense of a quality. The *pañca* in *pañca janāḥ* is a peculiar adjective quite different from *śveta* ('white) in *śvetaḥ asvaḥ* ('white horse'). The number-words are not *guṇa-viśeṣaṇas* ('qualitative qualifiers') they are, therefore, called as *saṁkhyā-viśeṣaṇas* ('numerical qualifiers') by Pāṇini and others.

9

Types of Mathematical Operations

According to ancient Indian mathematicians, there are eight main types of operations to be performed in mathematics, which are basic. Bhāskarācārya calls them as *parikarmāṣṭaka,* or *aṣṭa parikarmāṇi.* The technical term for 'mathematical operation' used by the ancient Indian mathematicians is either *parikarma* or *parikriyā* or simply *kriyā; cf.* Bhāskarācārya in his mathematical text *Līlāvātī*: *atha bhinnaparikarmāṣṭakam* (p. 22 etc.). The eight main mathematical operations which are enumerated in *Līlāvatī*[24] are: *saṁkalita* (addition or summation), *vyavakalita* (subtraction), *guṇana* (multiplication) also called *hanana, bhāgahāra* (division), *kṛti* or *kṛti-karaṇa* (squaring), *kṛti-mūla - karaṇa* (finding out the square-root), *ghana-karaṇa* (finding out the cube) and *ghana-mūla - karaṇa* (finding out the cube-root). Besides the above terms which are strictly technical, other words are also used for indicating the different operations; for example, √*yuj* or its derivative *yoga* with or without *sam* signifies 'addition'; the word *dhana* is also used to signify the same. The root *sam* + √*yu* and its derivative *sam+ yuta* also expresses summation. Even the root √*i* with *anu* and its derivative *anu-ita* or even the root *i* with the *upasargas sam* + *ā* and its derivative *sam+ā+ita* (= *sameta*) is also used. In the case of subtraction, besides *vyāvakalana*, words, like *vi+√yu* or *vi* + *√yuj* and their derivatives *viyuta* or *viyoga* are also used; moreover, a word like *ṛṇa* also indicates subtraction. For multiplication, *√guṇ* the *√han* and their derivatives are used. For division, *bhaj* with or without *sam* or *vi* and its derivaties like *vibhāga, sambhāga* are also used; 'division' is also indicated by the root *√hṛ* or its derivatives like *hara* or *hāra*. These are the technical terms for the four main mathemaical operations used in the classical stage of Sanskrit. The terms for the remaining four mathematical operations have remained the same throughout, right from the times of Aryabhaṭa I to those of Keshava of the fifteenth century AD.

We have no glossary of technical terms for the mathematical operations given in the Vedas. We have, therefore, no alternative but to conjecture the technical terms on the basis of the texts in which some mathematical operation is seen to have been given, suggested or implied. The following words for the different mathematical operations have been found out as suggestive of the said operations after a very close and exhaustive study of the context and situations given in the Vedas.

10

Signs and Sign-Words for Mathematical Operations

If we go through any modern book on mathematics, we find that every page of it is sprinkled with different signs indicating different mathematical operations, such as + (plus) used to indicate the addition, −(minus) for substraction, X for multiplication, ÷ for divisonn, √for roots etc. Although the Veda is not a book exclusively dealing with mathematics it does contain, as we have seen and soon we will see, certain mathematical data. As such, it does require the help of some signs which will show the different mathematical operations between the different numbers, or in general, mathematical entities.

But the difficulty in the case of Vedic literature in searching for the mathematical operational signs is that it was never a written document in old times and secondly that even if it is printed now, it is not printed in the mathematical form. Again, the whole literature, as the Sanskrit tradition goes, has been handed down orally through the many generations. Moreover, whatever mathematical data it contains, the whole data are presented not in the form of signs but in the form of words.

To explain, let us take the example of a mathematical expression like 2+3=5. The expression, in the form in which it is given here, contains all signs or symbols and no words; the words are spoken by the speakers. In the form of words, the expression can be expressed in any one of the following ways: "(i) Two plus three is equal to five; or (ii) Two added to three become five, or (iii) three added to two give out five" etc. The Vedas have come down to us in exactly the latter way, i.e. in the form of a language. In such circumstances, the search for finding out the signs indicating the different mathematical operations becomes very difficult and is hindered at every step. It must be remembered here, therefore, that whatever we will find in the Vedas will be only in the form of words or language which are spoken and not in the form of signs which are written.

10.1. Sign-words for addition:

Let us start at the initial stage from the mathematical operation of addition which is the most primary and the simplest one for any primitive society to perform. The examples of addition from the RV. are as follows:

10.1.1 *use of 'ca' for addition*

RV. 1.32.14: *nava ca navatim ca* states the addition of nine and ninety and gives us the number *nava-navati* i.e. ninety-nine. While stating this, the text uses the word '*ca*' meaning 'and' as the sign-word. In modern signs , the statement of addition can be written down as 9+90=99.

It should be noted that Pāṇini also uses the word '*ca*' for compounding two words in *dvandva-samāsa* in the *sūtra, cārthe dvandvaḥ,* 2.2.29.[25]

10.1.2: *use of sākam for addition*

RV. 4.26.3: *nava sākam navatīḥ,* gives the addition of *nava* (9) and *navati* (90) which is *nava-navati* (99). There are many passages in the RV. itself in which the use of *sākam,* which means '*together*' (from √*sac* 'to be together') gives the process of 'addition'.

If we examine the occurrences of the word *sākam* in the RV., we find that the word is used both with the number-words as well as with non-number words (like *āgni, sūrya, raśmi* etc). In the case of the use with the non-number words, it means just 'together with, or jointly' etc. In the case of the use of the word *sākam* with the number-words, we find it is used in three cases, viz. the nom., acc., and the instr. In all these cases, the word signifies the meaning of 'addition' if the number of the number-words is more than one; *cf.* RV. 1.164.48, 4.26.3.7; 6.27.6, 7.99.5, 8.86.14 etc. It is also to be noted that the use of the acc. exceeds that of the nom. and instr. when the meaning of 'addition' is signified. Use of the instr. is found in greater proportion in the case of the non-number words and means only 'together with, jointly' etc. This same meaning is given out when the nom. case is used.

It is also to be noted that though the words *sākam* and *saha* are rough semantic equivalents of each other, the word *saha* is never used in the contexts of number-words. We, therefore, cannot equate *saha* with the meaning of 'addition, summation' etc. In other words, while *sākam* seems to be a mathematical technical term, *saha* does not seem to be so intended. In the three Ṛgvedic passages, viz. 5.62.1 (*daśa śatā saha tasthuḥ tad ekam),* 5.62.6 (*sahasrasthūṇam bibhṛtaḥ saha dvau),* and 8.29.8 (*vibhir dvā carataḥ ekayā saha*) where *saha* occurs together with the number-word it means 'together with' and is used adverbially going with the verbs and does not mean the operation of 'addition'.

There is yet another important point which deserves to be noted. When the Vedic poets intend addition of the numbers, they invariably use the sign-words *ca* and/or *sākam.* But when these two sign-words are not used, what is intended is not the summation but multiplication. Thus, the phrase *nvatim sahasrā,* RV. 10.98.11 intends not the sum of 90 and 1000 (i.e. 1090) but the multiplication of 90 and 1000 i.e. 90,000; so also, whereas the passages *nava ca navatim ca* and *nava sākam navatīḥ* intend the summation, the passage *navānām navatīnām*, RV. 1.191.13 implies not the summation but the multiplication i.e. nine times ninety or ninety times nine i.e. 9x90=810. The absence of *ca* or *sākam* in the latter example is specially noteworthy. The words *nava* and *navati* in RV. 1.191.13 are in regular relation of adjective and substantive (*viśeṣana-viśeṣya-bhāva*), while they are both substantives in RV. 1.32.14 and 4.26.3; *cf.* also other examples like *sahasrāni śatā* (RV. 4.30.15.)= thousand hundred or hundred thousand (i.e. 1000,00 or 100,000). As the words *ca* and *sākam* are absent, the phrase does not mean 1000+100=1100.

It will be clear from the above discussion that both the words *ca* and *sākam* can stand and be taken as verbal equivalents of the mathematical sign + (plus) used in modern times.

10.2. Sign-words for subtraction

Actually, the knowledge of the mathematical operation of addition authomatically gives rise to the knowledge of the operation of subtraction, since the two processes—or, rather thought—processes—are opposite to each other. Also, conversely, the knowledge of subtraction-operation implies the knowledge of addition.

Yet, surpprisingly enough although all the Vedic texts exhibit full knowledge of the concept of subtraction, none of them explicitly mentions the process by any word. Unlike in the process of addition noted above, no sign-word is available in the case of the process of subtraction. The knowledge of the process of subtraction can only be inferred by the actual examles; *cf.* for example the Ṛgvedic passage, 1.164.45: *catvāri vāk parimitā padāni ..... guhā trīni nihitā neṅgayanti turīyaṁ vāco manuṣyā vadanti,* in which the remainder 'one' is accounted for when 'three' is subtracted from 'four'; for other examples, see below the section on 'the examples of subtraction'. In all these examples, however, we donot find any sign-word for subtraction used by the Vedic poets.

We may however, cite one, single solitary example of a Vedic passage in which we find the word *avama* (= 'lower, less than') seems to have been used as the sign-word for 'subtraction'. The passage in question is from AV. and is as follows: AV. 19.47.4 and 5: .... *catvāraś catvāriṁśac trayastriṁśac ca vājini* (4 *cd*) *dvau ca te viṁśatiś ca te rātryekādaśāvamāḥ* (5 *ab*)

The context is: the author is counting the nights; he starts from the number 99 (*navatir nava,* AV. 19.47.3) and by deducing or subtracting the number *ekādaśa* (= 11) from 99 states the remaining number as 88. He again subtracts 11 from the remainders and comes upto 22 (*dvauca viṁśatis ca*). He then remarks that the number 22 is arrived at as it is "less by elevan" than the number 33 mentioned is 4 *cd* quoted above. The *pāda* 5 *b* may also mean that the number 11 is "less by eleven" than the number 22.

Yet, though the interpretation of the word *avama* as the sign-word for "subtraction" seems to be ambiguious and not very much convincing, it does certainly seem to have the function of a sign-word for "subtraction".

In line with the word *avama,* we can also mention words like *avara, uttara, uttama, adhara, adhama* which in numerical contexts can mean 'greater than or less than', and can be used as 'sign-words' in substraction. Yet, primarily these words seem to be from special context and not a numerical one.

The other word which can be cited as used as a sign for subtraction and which is found in later literature also is *ūna,* from √*un* (*cf.* Pāṇinian *dhātupāṭha,* ūna *parihāṇe*). It is used in RV. 1.53.3, repeated in AV. 10.21.3 (*mā tvāvato jarituḥ kāmam ūnayīḥ*); the form *ūnayīḥ,* irregular for *ūnayaḥ,* (Aor. 2nd sing.) is explained by Pāṇini in the *sūtra,* 3.1.51 and is a hapax. in the Vedic literature. It is this root √*ūn* which gives out the derivative *ūna* meaning 'less than': The glaring examples of the word *ūna* signifying subtraction are the number-words *eka-ūna-viṁśati* (for 19), *eka-ūna-triṁśat* (29) etc; TS. 7.4.7, as we have noted earlier (*cf.*

Table No. 1) notes the number-words for 49 as *ekasmān-na-pañcāśat* (= fifty minus one); or even *ekasyai-na-pañcāśat* (= one remaining for fifty); *cf.* TS. 7.4.7: *sa ekasmānnapañcāśam apaśyat ...... ūnātiriktā vā etā rātrayaḥ/ūnās tad yad ekasyai na pañcāśat/* The same word occurs in AV. also (*cf.* AV. 10.8.44 etc) in the same sense. AV. 10.8.15 clearly gives the meaning of subtraction as : *dūre ūnena hīyate.* The word *pūrṇa* 'full' is contrasted in meaning with the word *ūna* 'less, deficient' etc. Later classical Sanskrit uses the word *ny-ūna* (*ni + ūna*) for the Vedic *ūna* more frequently. The word is used in Veda in a general, non-numerical context always, but may signify the meaning 'subtracted from, less than' in numerical context. The general, non-numerical meaning in the Veda is 'deficient, lacking, incomplete' etc. *cf.* VS. 3.17: *tanvā ūnam.* KS. 21.3 and 23.1 equates the word with the word *chidra,* 'hole, hollow' etc; *cf.* KS. 4.3 *yadevāsya ūnam yac chidram,* and KS. 23.1: *ūnam iva vā etac chidram iva ..... tadevāpūrayatiachidratvāya; cf.* in this connection the compounds *acchidroti* (= *acchidra+ūti*) found in the RV. 1.45.3 and *acchidra–ūdhnī* found in RV. 10.133.7 in which *acchidra* = 'not less, not deficient' i.e. 'full'.

10.3. Sign-words for multiplication—multiplicatives

Just as every language contains words for cardinal and ordinal numbers, it also contains words which are used as multiplicatives or which indicate the meaning of 'repetitions so many times'. We have, for example, the English words for number 1 as 'one' (cardinal), 'first' (ordinal); we have also the word 'once', which signifies the meaning of 'repetition one time'; so also we have 'two', 'second' and 'twice' and so on. The Vedic language also contains and notes many words which signify the sense of 'repetition so many times' and suggest the knowledge of the mathematical process of multiplication. All such multiplicative words are derived from their cardinal bases which are easily decipherable.

10.3.1. *Multiplicative from 'eka'*

The general multiplicative is formed with the word *vṛt* which is derived from √*vṛ*, 'to cover'; we have, therefore, the words like *eka-vṛt, dvi-vṛt, tri-vṛt* etc. This word is available only for the first three number-words viz. *eka, dvi* and *tri*; for *eka-vṛt* meaning 'one covering i.e. time, 'once', *cf.* TS. 5.2.3: *ekavṛd eva svargam lokam eti*' which is repeated in KS. 20, KKS. 31.3. The Vedic word *vṛt* from *vṛ* is substituted by *vāra* (also from *vṛ*) in later classical Sanskrit; the use of *-vāra* with number-words in the sense of 'so many times' is not available in the Vedic Saṁhitās.

Besides the word *-vṛt*, the *kṛdanta* - form of kṛ with zero-suffix, viz. - *kṛt* is also used with the word *eka* only. And we have the form 'eka-kṛt' meaning 'once'. But one peculiar change to be noted is that while compounding with the word - *kṛt*, the word *eka* undergoes a substitution by *sa-* and we have the form for the meaning 'once' as *sa-kṛt*. The Sanskrit *sa* goes back to the IE. base **sem*.[26] We have, therefore, two forms of the multiplicative from *eka*, viz. *ekavṛt* and *sakṛt*.

Pāṇini explains the form *sakṛt* by the substitution of *eka* by the entire morpheme *sakṛt; cf.* the *sūtra, ekasya sakṛt ca* , 5.4.19. The word *sakṛt* which seems older than *eka-vṛt* is available right from the times of the oldest *saṁhitā, viz. RV; cf.* RV. 1.105.18 etc. Though Pāṇini states total substitution for *eka* by *sakṛt*, from linguistic point of view, the real substitution seems to be of *eka* by *sa*, because we get a form like *sa-vṛt* meaning 'once' in KS. 17.7 (*savṛdasi savṛte tvā*); also we get forms like *dvi-vṛt* and *tri-vṛt* for which see below.

10.3.2. *Multiplicative from dvi*

As in the case of the word *eka*, so in the case of the word *dvi* also, the word *vṛt* is appended to *dvi* and we have the form *dvi-vṛt* meaning 'two times i.e. twice'; *cf.* KS. 11.4: *dvivṛt hiraṇyam dakṣiṇā*.

Besides the form *dvi-vṛt*, we find another device used by the Vedas suggesting multiplication by the number 'two'. The suffix attached with *dvi* in this case is *s* i.e. *visarga*. Pāṇini notes this phenomenon in the *sūtra, dvi-tri-caturbhyaḥ suc*, 5.4.18. The Pāṇinian suffix is *suc* with *u* and *c* both zeroed. And we have,

dvi+suc

= *dvi+s* (*u*, *c* = o)

= *dvis*

= *dviḥ*

We have thus two mulplicative forms from the number-word *dvi* as *dvi-vṛt* and *dviḥ*. Out of this, the form *dvi-vṛt* occurs only once in the whole of the Vedic literature comprising the nine saṁhitās taken here for study, and that too, in a later *saṁhitā* like Kāṭhaka noted above. The multiplicative *dviḥ,* however, occurs right from the times of the oldest *saṁhitā* of the RV. to the end of the period of classical literature. The corresponding forms for *dvih* in other languages are: Av. *bis*, GK. *bis,* Lat. *bis*, Old Lat. *duis*, Goth. *tvis; cf. K. Brugmann* , *ibid*. III.48; for occurrences, *cf.* RV. 1.53.9 etc.

10.3.3. *Multiplicative from tri*

Like the previous two number-words viz. *eka* and *dvi*, the number-word *tri* also exhibits two forms of multiplicative, viz. one with *vṛt* as *tri-vṛt* and the other with the suffix -*s* as *tris* = *triḥ*. Both the forms are obtained right from the oldest Vedic stage. For *tri-vṛt*, *cf.* RV. 1.140.2 etc; for *triḥ*, *cf.*, RV. 1.20.7 etc. The corresponding forms in other IE languages are: Av. *thris, GK. tris,* Lat *ter*, "perhaps for *ters*", O. Ir. *tress*; *cf. K. Brugmann, ibid.* III. 48.

10.3.4. Multiplicative from *catur*

So far as the word *catur* is concerned, it exhibits its multiplicative formation only in -*s* and not in -*vṛt*; even in the case of -*s* also, the formation is available only in later *saṁhitās; cf.* TS. 2.6.7. (*catur upa hvayate*) MS. 1.6.8; KS. 6.4 and KKS. 4.3. It is to be noted that the multiplicative formation from *catur* with any of the two suffixes, viz. -*vṛt* and -*s* is not available in the four main *saṁhitās*, viz. RV; VS. SV and AV. For corresponding forms, *cf.* Av. *cathruś*, Lat. *quater*; *cf.* *K. Brugmann* , *ibid.* III. 48 f.

10.3.5: *Multiplicatives from number-words after catur*

In the case of the number-words after *catur*, however, neither the word -*vṛt* nor the suffix -*s* seems to have been used. What is done is that a suffix -*kṛtvaḥ* is attached to the number-word and the multiplicative is formed. The suffix *kṛtvaḥ* is also not found in the earlier four main *saṁhitās*, viz. RV., VS, SV and AV; It is found only in the later *saṁhitās*. The number-words to which the multiplicative suffix *kṛtvaḥ* is applied are *tri*, *pañca*, *saṭ*, *aṣṭa*, *nava*, *daśa*, *ekādaśa* and *dvādaśa*; *cf.* for *tri*, MS. 4.1.10: *triḥ kṛtvaḥ*; for *pañca*, TS. 6.1.1, 9.5: *pañcakṛtvaḥ*; *for ṣaṭ*, TS. 6.5.3: *ṣaṭkṛtvaḥ āha*; for *aṣṭa*, TS. 6.4.5: *aṣṭau kṛtvaḥ*; for *nava*, MS. 4.5.7. *nava kṛtvaḥ*; for *daśa*, MS. 3.7.4: *daśa kṛtvaḥ*; for *ekādaśa*, TS 6.4.5: *ekādaśa kṛtvaḥ* and for *dvādaśa*, also TS. 6.4.5. In the case of the word *tri*, a peculiar fact notable is that both the suffixes viz. -*s* as well as -*kṛtraḥ*, are applied to it to give out a kind of double multiplicative as *triḥ kṛtvaḥ* (see above). The other thing to be noted in the case of the suffix -*kṛtvaḥ* is that as its independent accent shows, it seems to be an independent word and not a suffix; thus in *pañca kṛtvaḥ* (TS. 6.1.1), both the words *pañca* and *kṛtvaḥ* are accented. Does it show that the morpheme -*kṛtvaḥ* was originally an independent word and not a suffix as is thought by grammarians like Pāṇīni and others?

Pāṇini describes the multiplicative forms from number-words by applying the suffix -*kṛtvaḥ* i.e. *kṛtvasuc* (in Pāṇinian technical reconstruction). The *sūtra* is: *saṁkhyāyāḥ kriyābhyāvṛtti-gaṇane kṛtvasuc*, 5.4.17. The word *kriyābhyāvṛttigaṇana* in the *sūtra*, for expressing the significance of 'repetitions of actions' (*kriyābhyāvṛtti*) is notable; the word *abhyāvṛtti* from √*vṛt* 'to repeat' is a synonym of the suffix/word *vṛt* noted above.

For other equivalents, *cf.* Lith. *uenaṁ karat* ('once'), *du kartu* ('twice'), *tris kartus* ('three times') etc. *K. Brugmann* (*ibid.* III. 49) connects the present *kṛtvas* (from *kṛ*) with *-kṛt* (also from *kṛ*) in *sakṛt* (= once) discussed above.

Besides the suffix *kṛtvas* and *s,* Pāṇini also notes a suffix *-dhā* from only the word *bahu* in the sense of *kriyābhyāvṛttigaṇana;* it is applied optionally. The *sūtra* is: *vibhāṣā bahor dhā aviprakṛṣṭakāle*, 5.4.20. Yet, the suffix *-dhā* is seen to be used in a distributive sense (i.e. 'in so many ways') and not in a multiplicative sense in all the Vedic *saṁhitās; cf.* the famous Ṛgvedic passage. *ekaṁ sad viprā bahudhā vadanti* , 1.164.41.

In classical Sanskrit, the sense of a multiplicative is signified, besides all the above three suffixes viz. *vṛt, s* and *kṛtvas,* by yet another derivative from √*vṛ* 'to cover'; the derivative is *-vāra* as in *eka-vāra, dvi-vāra, tri-vāra* etc. Yet this suffix seems to be absent in the Veda. Perhaps, can we say that the suffix is found in forms like *bhūri-vāra* and *puru-vāra* available in the RV. also? *puru* and *bhūri* can be taken as number-significands without number-words. But the difficulty is that these two forms are adjectives and not adverbs.

10.4. Sign-Words for division–distributives

Just as the operation of addition automatically gives rise to or implies the operation of subtraction, the operation of multiplication leads automatically to its reverse operation of division. In multiplication, two numbers or multiplicands give out a single number; in division the result, subjected to a process, gives out its multiplicands.

In the Vedic literature, we do not get a definite, concrete evidence, in the context of numbers, for the knowledge of the process of division on the part of the Vedic poets. Yet, that they knew the division of an entity, which may not be necessarily numbers, into equal parts is corroborated by certain but few passages. The words used for indicating the division are: √*bhaj* with or without the *upasargas vi* and *sam* and its derivatives like *bhāga* etc; √*vic* 'to separate'; √*san* and √*van*, both meaning 'to divide or distribute equally'. (*cf.* Pāṇinian *dhātupāṭha: vana ṣaṇa sambhaktau*; note the *upasarga sam*, which means 'equal', in Pāṇini's wording). In RV. 1.81.6 (*vi bhajā bhūri te vasu*), Indra is invoked to divide or distribute' his abundant wealth; In RV. 1.27.5 (ā no bhaja parameṣu... madhyameṣu...) Agni is requested to 'distribute or divide' his *vasu* (= riches) into three parts *parama, madhyama* and the last called *antama*, which the last, he is invoked to give to the devotees (*cf.* 1.27.5*c: śikṣā vasvo antamasya*); *cf.* also RV. 1.162.4 (*trir mānuṣāḥ pary aśvam nayanti atrā pūṣṇo prathamo bhāgaḥ* ....) in which the first part out of three parts is said to belong to Pūṣan. Incidentally this implies the knowledge of the fractions also. For the use of *vic, cf.* RV - 10.124.5. Division or distribution of *vasu, rāyaḥ, dhanam* etc. is a favourite idea of the Vedic poets; *cf.* RV. 1. 123.4 etc. In RV. 3.30.7 (*abhaktam bhajate*) the deity is said to divide that which is undivided.

So also are the roots *van, san* used to signify 'division or distribution'.

There is yet a difference discernible in the usage of the root *bhaj on the one hand and the roots san* and *van* on the other. While the root *bhaj* is used in the sense of a regular 'division into parts', the roots *san, van* signify only the 'distribution'; *cf.* the oft-quoted phrase *vājam* or *svaḥ san/van* in the RV. Also, whereas the root *bhaj* is used in the division of concrete things like *haviṣ, vasu* etc. the roots *van, san* are used in the context of division of abstract things like *vāja* (strength), *svaḥ* (light) etc; *cf.* the compounds *vāja-sāḥ, svar-ṣāḥ* in the RV.

As for √hṛ or its derivatives like *hara* which are used to signify 'division, divisor or denominator' in later mathematical texts of Āryabhaṭa, Bhāskarācārya and others, the Ṛgvedic passage, 10.162.4 (*yas ta ūrū viharaty antarā dampatī śaye*) may serve as the oldest evidence for the later usage of √*hṛ* in the above-mentioned sense.

10.4.1. *The distributives in -dhā*

Though we do not get for certain any verb or word which would signify the division, we do get distributive adverbials which signify the division of something, esp. numbers into 'so many equal parts.' These adverbials are formed by applying the suffix *-dhā* or *-dhātu* to the number-words. And we have the forms in *-dhā* from number-words, like *ekadhā* (= in one way, fold etc.) *dvidhā* (= in two equal ways), *tridhā* or *tridhātu* (= in three equal ways) etc.

The Pāṇinian *sūtra* which prescribes the suffix *-dhā* is: *saṁkhyāyāḥ vidhārthe dhā*, 5.3.42. BD's commentary on this makes it very clear that the suffix is applied only in the sense of "the way any action is done"; *cf.* BD: *kriyāprakārārthe,* which emphasizes the "ways in which the action is done". The next *sūtra*, *adhikaraṇavicāre ca*, 5.3.43 makes it very clear. BD's remark on this *sūtra* is also very bold; *cf.* BD: *dravyasya saṁkhāntarāpādāne, saṁkhyāyāḥ dhā syāt,* which means "*dhā* is used to get other *saṁkhyās* i.e. numbers from a single *saṁkhyā*." His example, *ekam rāśim pañcadhā kuru* (= make into five-fold one single number) has an implied sense of 'equality' in division. Kāśikā also makes the things very clear when it declares that *dhā* is applied only in the context of any action with reference to a number; *cf.* Kāśikā on 5.3.42: *vidhā prakāraḥ; sa ca sarvakriyāviṣaya eva gṛhyate.* The word *vicāra* in the *sūtra*, 5.3.43 means, according to kāśika, 'obtaining another number' from a single number; *cf. vicāraḥ saṁkhyāntarāpādānam-ekasya anekīkaraṇam, anekasya vā ekīkaraṇam.* The wording *ekasya anekīkaraṇam* (= making or transforming one entity i.e. number into many) expresses in clear terms the process of division; *cf.* the examples, *ekam rāśim pañcadhā kuru, aṣṭadhā kuru.* Kātyāyana in his *vārttika* on the Pāṇinian *sūtra* explains *vidhārthe* as: *dhāvidhānam dhātvarthapṛthagbhāve*; Patañjali explains *vidhārtha* as: *kas tarhi dhātvarthapṛthagbhāvaḥ? kārakāṇām pravṛttiviśeṣaḥ kriyā; kriyāprakāre ayam bhavati; vidhayuktagatāś ca prakāre bhavanti* . All this means that although the meaning of 'equality' may not be present, the sense of 'division' is implied.

And we have the formations in *-dhā* from number-words as *ekadhā, dvidhā, tridhā* etc., meaning 'division of an entity into one, two, three etc. ways'. In numerical context, it amounts to mathematical operation of division.

In the case of the number-words *eka, dvi* and *tri*, however, we have also the optional forms as *aikadhyam*, *dvedhā* and *dvaidham*, and *tredhā* and *traidham*. For *aikadhyam*, the *sūtra* is: *ekād dho dhyamuñ anyatarasyām*, 5.3.44; for *dvaidha* and *traidha*, the *sūtra* is: *dvi-tryoś ca dhamuñ*, 5.3.45. Thus, in the case of *eka*, the suffix is *dhya*, while in the case *dvi* and *tri*, the suffix is *dha*. Both the suffixes ending in *ñ* bring about the *vṛddhi* of the initial vowel of the respective words to which they are attached; *cf.* for *vṛddhi* the Pāṇinian *sūtra, taddhiteṣv acām ādeḥ* , 7.2.117. The process is:

eka+dhyamuñ

= *aika+dhyamuñ*

= *aika+dhya*

= *aikadhya*, which in meaning is the same as *ekadhā*.

For *dvaidha* and *traidha*, the process is:

dvi/tri+dhamuñ

= *dvai/trai+dhamuñ*

= *dvai/trai+dha*

= *dvaidha/traidha*, which in meaning are the same as *dvidhā/tridhā*.

For the other optional forms *dvedhā* and *tredhā* from *dvi* and *tri* respectively, the Pāṇinian *sūtra* is: *edhāc ca*, 5.3.46, which lays down the substitute *edhā* for *dhā*; and we have,

dvi/tri+dhā

= *dvi/tri+edhā*

= *dv/tr+edhā* (the final *i* = O by *ṭeḥ*, 6.4. 155)

= *dvedhā/tredhā*

So we have the distributive adverbials from *eka, dvi, tri* as *ekadhā/aikadhyam, dvidhā/dvedhā/dvaidha* and *tridhā/ tredhā/traidha* respectively. For all other number-words from *catur* onwards, *dhā* is the only suffix; they have no other optional forms. And we have *caturdhā, pañcadhā* etc. From the number-word *ṣaṣ* (= 6), we have the form *ṣoḍhā*; the process is:

ṣaṣ+dhā

= *ṣa-u+dhā (ṣ>u* according to *Vārttika* quoted above,

= *ṣo+dhā* (a+u = O)

= *ṣo+ḍhā* (dh>ḍh)

= *ṣoḍhā.*

The following distributive adverbial forms are attested in the nine Vedic texts taken here for study: *ekadhā, dvidhā/dvaidha, tridhā/tredhā, caturdhā, pañcadhā, ṣoḍhā, saptadhā, aṣṭadhā, navadhā, daśadhā* and *dvādaśadhā*. And correspondingly we have the division of a number by all the above numbers. From non-number-words having numerical significance, we have *bahudhā* (from *bahu*) and *purudhā* (from *puru)*, both meaning 'many, plenty of' etc. 10.4.2.

Distributives in -śas:

Besides the distributives in *-dhā*, we have also another type of distributive found in Sanskrit. These are formed by applying the suffix *-śaḥ*, Pāṇīnian *śas*; and we have the forms as *ekaśaḥ* etc. But they are not available in any of the nine Vedic *samhitās*; *cf.* K. *Brugmann, ibid.* III. 51.

11

The Concept of Sets or Groups

Besides the words which lead us to assume the knowledge of the four main mathematical operations on the part of the Vedic people, we have also a set of words which shows that they had also the knowledge of counting by sets or grouping things together. This type of counting by groups is found in some primitive tribes all over the world, like the Weddas of Srilanka or Bakairi of South America.[27] What is involved in this technique is that the many things which seem apparently difficult and boring to count are piled up in groups or sets of twos, threes etc; and then the sets are counted and the final number of sets is multiplied by the number of things contained in each set. Thus, 'two fours' or 'four twos' give us the number 'eight' (8). The Vedic people also resorted to this method many times. Actually, the number-words for all the infinite numbers can also be taken as 'the words for groups'; to illustrate, '*daśa* (10)' can be taken as the number-word for a group of 'ten' things. The device of multiplication, as can be seen from the previous discussion on the topic, is a very good example of indicating the working of this method. When, for example, the

Vedic poets use the phrase '*dviḥ daśa*' for 20, what they do is actually that they repeat the group of *daśa* 'two times' (*dviḥ*).

But besides this, the Vedic people had separate words for the different groups. These words are formed by applying the suffix -*taya* to the number-words excepting the word *eka* for 'one'. The Pāṇinian *sūtra* which lays down the suffix -*taya* is *saṁkhyāyāḥ avayave tayap*, 5.2.42. Thus we have *dvi+taya* = *dvitaya*, 'a group which has two members'; *tri+taya* = *tritaya*, 'a group which has three members' etc. The word *avayava* means 'factor, multiple' etc.

In the case of the words *dvi* and *tri*, we have also the optional forms as *dvaya* and *traya*, the suffix in this case being '-*aya*' (Pāṇinian -*ayac*) according to the Pāṇinian *sūtra*, *dvitribhyām tayasyāyaj vā*, 5.2.43. The same suffix viz. -*ayac* is available in the case of the word *ubha* also, giving out the form *ubhaya*, *cf.* *ubhād udātto nityam*, 5.2.44. Macdonell calls these words as 'multiplicative adjectives'[28] Since they are adjectives, they are declinable and follow the gender, number and case of their substantives. For *catur*, we have the form *catur-vayam* in RV. only (1.110.3; 4.36.4). The other form with -*taya* as *catuṣṭaya* is not attested in the Vedic *saṁhitās*, except in *Śaunaka* (10.2.3) and *Paippalāda* (16.59.3); it is, however, found in classical Sanskrit. We, therefore, get only the forms *dvaya*, *traya* and *daśa-taya* from *dvi*, *tri* and *daśa* respectively; we have also the form *ubhaya* in the Veda from *ubha*.

Besides the suffix -*taya*/-*aya*, we have in the Veda also the suffix -*ka* in the sense of 'group'. The only forms that are available with this suffix are *ekaka* (from *eka*) *dvaka* (and not *dvika*, from *dvi*) and *trika* (from *tri*); *cf.* RV. 10.59.9.

Besides the words which signify the specification of the number of members in a group or set, we have also the general words for groups. They are *gaṇa* and *vrāta*. The *gaṇa* is of the *devas* (RV.4.35.3: *devānām gaṇa*) and of the Maruts (RV. 1.14.3 etc; *mārutam gaṇam*), of Indra, RV. 1.23.8 and of Bṛhaspati, AV. 20.88.3. The word *vrāta* meaning 'set' seems to have been used exclusively in the context of inanimate things such as the dices; *cf.* RV. 10.34.8: *tripañcāśaḥ krīḍati vrāta eṣām.* Note the word *tri-pañcāśa*, an ordinal from *tri-pañcāśat* (53) which specifies the number in the *vrāta* of the dices. Presently we have 52 cards in the set of the playing cards.

Incidentally, it is also very interesting to note that the technique of many *gaṇas* forming into a bigger *gaṇa*, called *mahāgaṇa* was also utilised by the Vedic people, perhaps for the sake of the convenience of counting and calculation. AV. 19.22.16 (*gaṇebhyaḥ svāhā*) and 19.22.17 (*mahāgaṇebhyaḥ svāhā*) offer oblations respectively to *gaṇas* and those deities 'who form the *mahāgaṇa*'. Indra seems to be the first and the oldest deity of *mahāgaṇa*. RV.1.23.8 (*indrajyeṣṭāḥ marudgaṇāḥ*) mentions Indra as the head or the eldest of the *gaṇas* (the plu. is notable) of Maruts. Thus Indra together with his own *gaṇa* and the *gaṇa* of the Maruts forms a *mahāgaṇa*. It is this idea of *gaṇas* classified under the banner of a *mahāgaṇa* which might have been at the root of the concept of the god *gaṇapati*, 'the ruler of the *gaṇas*.' It is to be noted that the word *mahāgaṇa* occurs only once here in the AV. Also, corrsponding to the idea of *mahāgaṇa*, as contrasted with simple *gaṇa, we have later the idea of mahā-gaṇapati* , 'the great(est) ruler of the *gaṇas*' (*mahā* to be connected semantically with *pati*) or 'the ruler of the great(est) *gaṇas*' (*mahā*, grammatically to be connected with *gaṇa*).

Besides the words *vrāta* and *gaṇa*, we have one more word, viz. *rāśi* (lit. heap) which connotes the idea of 'collection, collectivity, group or set, But the word is nowhere used in the mathematical contexts. It is used in the non-numerical context of *vasu* (*cf.* RV. 6.55.3: *vasoḥ rāśiḥ*), *go* (*cf.* RV. 9.87.9: *gonām rāśiḥ*) and independently in plu. as *rāśayaḥ*, referring perhaps to the heap of the crops or harvest in AV. 6.142.3: *akṣitāḥ santu rāśayaḥ.*

It is to be noted that the word *rāśi* signifies the 'collection of numbers' or 'mathematical expression' like $x^2+2xy+y^2$ in later mathematical literature. In astronomy, the word means 'the signs of the zodiac' like Aries, Taurus etc., which are nothing but the 'collection of constellations'.

12

Examples of Addition

We have seen before that concept of the mathematical operation of addition is fully known to the Vedas and exemplified it with examples of addition of small numbers below one hundred. Besides this, we have also seen that the whole number-system, which displays the principle of what modern mathematics calls as 'the arithmetic progression' is based on the simple principle of addition. Besides all this, we have also the evidence of the working of addition-technique in the case of the higher numbers i.e. those above one hundred. The present section collects all such examples in which the operation of addition in the case of higher numbers seems to be working.

An important point to be noted in this connection is that we do not know for certain that the Vedic people were using any number-symbols comparable to the modern ones like 1, 2, 3 etc. But that what they were using were number-words is certain. We also do not find any written symbols for the mathematical operations comparable to modern ones like +(plus), - (minus), x(multiplication), ÷ (division) and a host of others. Instead, as we have seen before, they were using words themselves for the mathematical symbols, like *ca, sākam* etc.

12.1. The best examples are provided by the number-words from *ekādaśa* onwards. We can deciphr the principle of addition underlying it.

12.2.2. Represented as a combination of 1+1

RV.1.95.1 describes Agni as having two forms; one is *hari* and the other is *śukra*; thus 2 is analysd as 1+1; *cf. dve virūpe carataḥ...harir anyasyām bhavati...śukro anyasyām; cf.* also RV. 1.164.20: *dvā suparṇā sayujā akhāyā...tayor anyaḥ pipplam svādv atti anaśnann anyo abhicākaśīti.* It should be noted that the word *anya*, which in later literature as well as in other contexts from the Veda itself, signifies the meaning of 'the other enemy' etc. means in the present contexts, which is interpreted mathematically here, 'one (of the two)'. Thus it also suggests the idea of parts, viz. 'out of.'

RV. 3.30.11 (*eko dve vasumatī samīcī indra ā paprau pṛthivīm uta dyām*) indicates only one entity viz. Indra by the word *ekaḥ* and two entities viz. *pṛthivī* and *dyaus* by the word *dve*; two (entities), therefore, are equal to *one*(*pṛthivī*)+one (dyaus). RV. 4.30.19 indicates the number two as a summation of 1+1; *cf. anu dvā...nayaḥ andham śroṇam ca;* thus *dvā* = one (*andha*)+one (*śroṇa*); *cf.* also RV. 9.86.42 (*dvā janā yātayan...narā ca śaṁsam daivyaṁ ca*) in which Soma is said to travel through two worlds, which are pinned down as *narāśaṁśa jana* and *daivya jana*.

It can be very well seen from the above passages that sign-word for indicating the operation of addition which is used is the word ca which means 'and'; in the form of a mathematical sign, we can write it down as + in the modern way. The particle *ca*, which is used as a conjunctive in the whole of Sanskrit literature can be mathematically interpreted to mean 'add, sum' etc. It is this meaning of *ca, viz.,* 'addition, combination' etc. which is perhaps intended by Pāṇini also when he framed the *sūtra, cārthe dvandvaḥ,* for defining the *dvandva*-compound.[29]

12.3.3. Represented as a combination of 1+1+1

As the following passages will show, what the Vedic people meant by 'three' was 'a combination of 1+1+1'.

RV. 1.13.9 (= RV. 5.5.8), which is an *Āprīsūkta*, mentions the number 3 with reference to the three independent deities as *ilā, sarasvatī* and *mahī*; *cf.* also RV. 3.4.8; 7.2.8 and 10.110.8. RV. 1.95.3 (*trīṇi jānā...samudre ekam divi ekam apsu ekam*) analyses 'three' as 1+1+1. RV. 4.58.4 states *tridhā hitam = indra ekam +sūrya ekam+venād ekam;* also RV. 10.185.1 refers to 3 as 1+1+1 (*cf. trīṇām......mitrasya+aryamṇaḥ+varuṇasya*). It is also to be noted that instead of using the number-word *eka* (= one), the Vedas sometimes list or mention the required entities and make up for the sum; thus in RV. 1.13.9; 5.5.8; 3.4.8; 7.2.8; 10.110.8; 10.185.1 etc. quoted above, the three entities are listed to make up for the sum 3. The VS-20.43 (*tisro devīh...sarasvatīḍā...bhāratī*) follows the same way.

RV. 1.164.44 (*trayaḥ keśinaḥ..., saṁvatsare vapate ekaḥ+viśvam ekaḥ abhi caṣṭe+ekasya dadṛśe na rūpam*) clearly explains 3 as 1+1+1. That the number 3 is an immediate consecutive of 2 is clearly inferrable from RV. 10.56.1 (*idam ta ekam para u ta ekam tṛtīyena jyotiṣā saṁ viśasva*) which uses the word *tṛtīya* (= third) after accounting two entities by the word *eka* repeated; thus 1+1 and then the next is *tṛtīya.* Here we have a clear consecutive sequence of 2 (explained as 1+1) and 3. *cf.* also RV. 10.48.7 (*abhi...eko abhi dvā kimu trayaḥ karomi* gives the sequence of the first three numbers as 1, 2, 3. Also, we have here the number 3 explained as 2+1. In RV. 3.2.9 (*tisro yahvasya samidhaḥ...tāsām ekām adadhuḥ...dve upa jāmim īyatuḥ*) we have 3 explained as the sum of 1+2. *cf.* also KS 21.1: *dvir dakṣiṇām āṁkte sakṛt savyam = trivṛd yajñaḥ*.

12.4. 4 as a combination of 3+1

RV. 1.164.45 (*catvāri vāk parimitā padāni...guhā trīṇi nihitā...turīyam...manuṣyā vadanti*) represents 4 as the sum of 3+1. Also, the number 4 is the immediate consecutive number of 3. Read with § 4.1.2 above, the sequence of the first numbers as given in the RV. is 1, 2, 3, 4.

12.5. 6 as a sum of 3+2+1

RV. 3.56.2 (*ṣaḍ bhārān...bibharti...tisro mahīḥ guhā dve...darśi ekā*) analyses the number 6 as the summation of 3, 2 and 1.

12.6 8 as a sum of 7 and 1

RV. 10.72.8 (*aṣṭau putrāso aditeḥ...devān upa prait saptabhiḥ/parā mārtāṇḍam āsyat*) states that the deity Aditi had eight sons; with seven, she went to gods; the remaining one was *mārtāṇḍa*. This clearly states that the number seven requires one more to arrive at the number eight. Thus, in figures, 7+1 = 8.

This can also be cited as an example of the operation of subtraction. When seven sons went to the gods, what remained out of eight was only one son. Hence, in figures, 8-7=1.

12.7. The number 10

The number 10 is arrived at by adding different numbers. MS. 3.3.3. obtains the number by adding 2 five times; *cf. dvyakṣaram loma, dvyakṣarā tvak, dvyakṣaraṁ māṁsam, dvyakṣaram asthi, dvyakṣaro majjā tad daśa; daśākṣarā virāṭ,* which is identical with KS. 21.4; KKS 31.19. It is to be noted that *virāj* = ten; *cf.* also MS. 3.3.7; 3.4.6 etc.

KS. 20.1 states that 10=5+5, *cf. pañca citayaḥ, pañca purīṣāṇi, tad daśa, daśākṣarā virāṭ; cf.* also TS. 7.5.8.4: *pañcabhis tiṣṭhantas stuvanti...pañcabhir āsīnāḥ; daśa sampadyate . cf. also* TS. 5.2.3.7; 5.6.10.3; 6.4.4.2 etc.

KKS 31.13 arrives at 10 by the addition of 9 with 1; *cf. nava vai puruṣe prāṇāḥ nābhir daśamī; cf.* also TS 7.5.15.2: *daśa havīṁṣi bhavanti, nava vai puruṣe prāṇāḥ, nābhir daśamī.*

12.8. The number 11

The number 11 is arrived at by the addition of two as well as three different numbers. In the former case, we have 11 = 10+1 and in the latter case we have 11 = 8+2+1; *cf.* for 11 = 10+1, MS 3.7.3; 9.3; KKS 41.2: *daśa vai paśoḥ prāṇāḥ, ātmā ekādaśa;* repeated in KS. 29.9 with *puruṣe* in the place of *paśoḥ* and in TS 6.3.10.5; 10.3.11.6 etc. *cf.* also KS. 28.3: *daśa vasavaḥ, indraḥ ekādaśa; daśa rudrāḥ indraḥ ekādaśa; daśa ādityāḥ indraḥ ekādaśa.* For 11 = 8+2+1, *cf.* KS. 26.4 *aṣṭā aśrayaḥ, dve paruṣī, ātmā ekādaśa cf.* also AV. 5.15.1: *ekā ca me daśa ca me; cf.* also MS. 4.6.2. *ekayā ca daśabhis'ca.*

12.9. The number 12

The number 12 is explained as an addition of 10+2, 6+6 as well as 2+2+1+1+4+2; for the former, cf. KS. 33.2: *daśa vai puruṣe prāṇāḥ, stanau dvādaśa.* For 12 = 6+6, *cf.* TS 7.3.11: *dvau ṣaḍahau bhavataḥ tāni dvādaśāhāni saṁpadyante; cf.* also TS. 5.6.10: *ṣaṭ citayo bhavanti, ṣaṭ purīṣāṇi, dvādaśa sampadyante.* For the third explanation of 12, *cf.* TS. 7.4.11.2: *dvādaśo vai puruṣaḥ* = *dve sakthyau, dvau bāhū, ātmā ca, śiraś ca, catvāry aṅgāni, stanau dvādaśau.*

12.10. The number 14

The number 14 is equal to either 7+7 or 10+4; for the former, *cf.* TS 7.3.3.4: *caturdaśarātro bhavati...sapta grāmyā oṣadhayaḥ sapta āraṇyāḥ.* For the latter, *cf.* TS 7.3.5.3: *caturdaśa etāḥ; tāsām yāḥ daśa...yāś ca catasraḥ diśaḥ...*

12.11. The number 15

The number 15 is equal to 10+5; *cf.* TS. 7.3.7.4: *pañcadaśa etāḥ; tāsām yāḥ daśa...yāḥ pañca* .

12.12. The number 17

MS 1.11.6 is repeated in KS. Both of them explain 17 as 4+2+1(=7)+10; *cf. saptadaśaḥ puruṣaḥ = catvāry aṅgāni, śirogrīvam, atmā, vāk saptamī, daśa prāṇāḥ.*

12.13. The number 18

TS 7.4.11.4 obtains 18 by adding 9 with 9; *cf. aṣṭādaśāhāni sampadyante* which is equal to *navāny anyāni, navāny anyāni.*

12.14. The number 19

The number word *navadaśa* (= 9+10) used by VS. 18.24, TS. 4.3.10.1 and KS. 17.4 etc. is enough to give us the idea of its derivation as 9+10. Yet, TS. 7.2.11.19 also gives 19 as 20-1; *cf.* the word *ekannaviṁśati* given by TS.

12.15. The number 20

TS. 7.3.7.4. obtains it in the following way: viṁśo vai puruṣaḥ - *daśa hastyāḥ aṅgulayaḥ, + daśa pādyāḥ aṅgulayaḥ,* which means 20 = 10+10. This passage is repeated many tims in almost all the later *saṁhitās*, as well as in TS. itself. KS. 20.13 explains 20 as equal to 'two *virāj*' *cf. yad viṃśatiḥ, dve virājau.* The word *virāj*, as we have noted earlier is equal to 10; *cf.* above *daśākṣarā virāj*; the number 20, therefore, is equal to 10+10, which are equivalent to 'two *virāj*'; *cf.* also TS 5.3.3.3: *yad viṁśatir, dve tena virājau.* It is to be noted that the number 20 is also designated by the number-word *saviṁśa; cf.* TS. 4.3.8.2 etc. Its derivation is not clear.

12.16. The number 21

RV. 7.18.11 states it as the sum of 1+20; *cf. ekaṁ ca yo viṁśatim ca*. MS. 3.6.3, 9.8; KKS 3.5.7 explain it as 10+10+1; *cf. daśa hastyāḥ daśa pādyāḥ aṁgulayaḥ, ātmā ekaviṁśaḥ; cf.* also TS. 6.1.1.8 repeated. TS. 7.3.10.5 explains it also as 12+5+3+1; *cf.* TS. *ekaviṁśatirātram āsiran, dvādaśa māsāḥ, pañca ṛtavaḥ, traya ime lokāḥ, asau ādityaḥ ekaviṁśaḥ* .

12.17. The number 22

AV explains it as 2+20; *cf.* AV. 5.15.2: *dve ca me viṁśatiś ca me; cf.* also AV. 19.47.3: *dvau ca te viṁśatiś ca; cf.* also MS. 4.6.2: *dvābhyām...viṁśatyā ca* .

12.18. The number 24

TS 7.4.11.4 derives the number 24 as 'four six-days', i.e. the number 6 added four times; *cf. catvāraḥ ṣaḍahāḥ bhavanti; tāni caturviṁśati sampadyante; cf.* also KS. 33.3: *catvāraḥ ṣaḍahāḥ bhavanti, tāni caturviṁśatir ahāni sampadyante.* MS 1.10.8 derives 24 = 2 *saṁvatsaras* which is = 2X12; *cf. yau dvau saṁvatsarau, tayoh caturviṁśatih.*

12.19. The number 25

KS. 33.8 explains the number 25 as 10+10+2+2+1; *cf. pañcaviṁśam stomam upayanti...daśa hastyā aṅgulayo daśa padyā dvau bāhū dve sakthyā ātmā pañcaviṁśaḥ.*

12.20. The number 27

It is explained as 'three nines' by the word '*tri-nava*', besides the word *saptaviṁśati; cf.* VS. 14.23: *ojas triṇavaḥ,* which Mahīdhara explains as: 24+2+1; *cf. Mahīdhara, caturviṁśatyardhamāsaḥ, dve ahorātre, saṁvatsaraḥ...triṇavaḥ;* he further explains *tri-nava* as *triguṇāḥ nava,* 'three-times nine'. The word *tri-nava* occurs many times in the different *saṁhitās*; *cf.* Kāṇvas. 15.7.2 etc; TS. 4.3.3.2 etc; MS. 2.7.20 etc; KS. 17.4 etc. KS. 34.9 equates *tri-nava* with *saptaviṁśati*; *cf.* KS 349: *saptaviṁśatir dīkṣeran, triṇavāyatanāḥ, triṇavā ime lokāḥ triṇavāyatanāḥ; cf.* also KS. 33.8: *sapta-viṁśatir grahītavyāḥ, triṇavāḥ ime lokāḥ.*

12.21. The number 29

The word *navaviṁśati* 29, explains it as the addition of 9+20; *cf.* TS. 7.2.11.20.

12.22 The number 30

KS. 33.3 explains it both ways, viz. 10+10+10 = 30 or '5 times 6'; *cf.* KS. 33.3 = *daśa hastyā aṅgulayo, daśa padyā, daśa prāṇāḥ tat triṁśat; cf.* also *pañca ṣaḍahāni bhavanti, tāni triṁśadahāni sampadyante.* MS. 1.10.8 obtains 30 by adding 9+1+9+9+2; *cf. nava hi prāṇāḥ, ātmā devatā, nava prayājāḥ, navānuyājāḥ, dvau ājyabhāgau, tat triṁśat.*

12.23. The number 36

KS. and KKS. arrive at 36 by adding 12+12+12; *cf.* KS. 20.1: *dvādaśa dakṣiṇataḥ, dvādaśa paścāt, dvādaśa uttarāttāt, tāḥ ṣaṭtriṁśat sampadyante* . MS. 1.10.8. states that three *saṁvatsaras* make 36 full months; *cf. ye vai trayaḥ saṁvatsarāḥ teṣām ṣaṭtriṁsat pūrṇamāsāḥ.* We know that the word *saṁvatsara* stands for one year of 12 months. Therefore, 36 = 3 *saṁvatsaras* = 3X12.

12.23. The number 33 = 3+30 or 30+3; also as 11+11+11

RV. 1.45.2 (*trayastriṁśatam ā vaha*) mentions the number 33 which is interpreted as the total of 3 and 30; *cf.* RV. 3.6.9: *triṁśatam trīṁs ca* ; also RV. 8.28.1: *ye triṁsati trayas paraḥ;* also RV. 8.30.2 = *ye stha trayas' ca triṁśac ca* ; *cf.* also VS. 20.36 (*tribhir...triṁśatā*).

RV. 1.34.11 (*ā nāsatyā tribhir ekādaśaiḥ...yātam*) explains 33 as 'three elevans' i.e. 11+11+11; this passage is repeated in VS. 34.47; *cf.* also RV. 1.139.11 = VS. 7.19 = TS. 1.4.10.1 The number 33 is explained in VS. 20.11 (*trayādevā...ekādaśa trayastrimsāḥ*) as the sum of three sets of 'elevans.' The plu. *trayāḥ* for *trayaḥ* is peculiar and deserves a notice. *cf.* also KS. 38.11.

12.24. Other numbers

Besides the above numbers, the numbers, 77, 99, 107 and 720 are given as combinations of the relevant numbers; thus 77 = 70+7; *cf. adhīn nv atra saptatiṁ ca sapta ca* , RV. 10.93.15; the number

107 = 100+7; *cf.* RV. 10.97.1. *śataṁ dhāmāni sapta ca* which is repeated in VS. 12.75; the number 99 = 9+90; *cf.* RV. 1.32.14: *nava ca yan navatim ca*; *cf.* also RV. 1.54.6; 84.13; 2.14.4;19.6; 4.26.3; 48.4; 5.29.6 etc; the number 720 as seven hundred added to twenty; *cf. sapta śatāni viṁśatiś ca tasthuḥ* (RV.1.164.11) which can be written down as 100+100+100+100+100+100+100+20. The number 360 is represented as the addition of 'three hundred and sixty'; *cf.* RV. 1.164.48: *tasmin sākam triśatā...śaṁkavaḥ arpitā ṣaṣṭiḥ* which can be written down as 100+100+100+60; *cf.* also TS 7.5.1.3: *trīṇi ca śatāni ṣaṣṭiś ca...saṁvatsarasya rātrayaḥ.*

A point deserves notice. If the numbers are written in symbols like 1, 2, 3 etc., we read them from left to right if a big number is given. Take, for example the number in figures like 12345; we read it from left to right as: twelve thousand three hundred forty five. That is to say, we start from the higher rank and go to the lower rank, step by step. But in the case of the Vedic phrases in words for numbers, the method of reading or stating has no definite order. For example, in *śatam ekam ca,* (for 101) or *śatam sapta ca* (for 107), the phrases are given in descending order; that is, they have started from the higher rank and then gone to lower rank; the same is the case with phrases like *sapta śatāni viṁśatiś ca* (for 720). Yet this order is not always abided by and we find the number is expressed in terms of any type of arrangement; that is to say, they may start with the lowest rank first and then go to the higher ranks, as in *ekam ca yo viṁśatim ca* , (for 21) or *nava ca......navatiṁ ca* (for 99) or *trayaś ca triṁśat ca* (for 33). We also get examples in which the mention of the number starts, neither from the highest nor from the lowest, but from the middle rank; *cf.* for example, *trīṇi śatā trī sahasrāṇi triṁśac ca nava ca* (for 3339). We thus find that the numbers are mentioned by their ranks; yet no order of ranks, either from lower to higher or from higher to lower, seems to have been followed strictly. But as we approach the period of classical Sanskrit, we find that the order from higher to lower ranks is strictly followed. Perhaps, the looseness of order of ranks in the Vedic period is possible because

(i) of the inflexional character of Sanskrit words, and

(ii) the numbers are stated, not in symbols, but in words. This, however, it must be remembered, does not permit us in any way to conclude hastily that writing was totally unknown to Vedic civilisation.

13

Examples of Subtraction

The passages quoted above for addition can themselves serve as examples of indicating the knowledge of the operation of subtraction on the part of the Vedic seers. Thus in RV. 1.95.1 and RV. 1.164.20, what remains is number 'one' when number 'one' is subtracted from number 'two'. RV. 1.164.45 gives us that when 'three' out of 'four' are accounted for or explicitly mentioned, what remains is 'one'. Similary what remains after 'seven' are taken out of 'eight' is 'one', which is accounted for in RV. 10.72.8 and 9 (*aṣṭau putrāso aditeḥ...devān upa prait saptabhiḥ, parā mārtāṇḍam āsyat*). Also RV. 3.2.9 (*tisro yahvasya samidhaḥ...etc.*) deducts 'one' from 'three' and the remainder is stated to be 'two' (*cf. tāsām ekām adadhuḥ...upa dve...īyatuḥ*).

KKS. 35.8. may serve as the best example of subtraction. It relates a story of the *Anuṣṭubh*-metre. It says that in the beginning the *Anuṣṭubh*-metre contained four *pādas*, with 'three' *pādas* of 'eight' syllables each and 'one' *pāda* of 'seven' syllables. Out of the 'seven' syllables of the fourth *pāda*, 'three' went to a *pāda* of 'eight' syllables; that became (8+3=) 'elevan'-and it was the *Triṣṭubh* metre. When 'three' of the 'seven' went away, what remained was 'four'; these 'four' went to the 'eight'–and that gave

(8+4=) 'twelve'–that was the *Jagatī*-metre. What remained was only one *pāda* of 'eight' syllables-and that was the *Gāyatrī* metre. To represent the literary statements by figures of numbers–

Anuṣṭubh = 8+8+8+7 (total 4 *pādas*)

Triṣṭubh is derived as 8+3 = 11, 3 being taken away from 7. Symbolically, Anuṣṭubh = {(8+3)+8+8+(7-3 i.e. 4)}

Jagatī is derived as 8+4, 4 being taken away from the fourth *pāda*; symbolically, Anuṣṭubh = {(8+3)+(8+4)+8}, in which the first (8+3) is *Triṣṭubh*, the second (8+4) is *Jagatī* and the last 8 is Gāyatrī. *cf.* KKS. 35.8:

sā eṣā anuṣṭup. tasyāḥ saptākṣaram ekam pādam aṣṭākṣarāṇi trīṇi. teṣām saptānām yāni trīṇi tany aṣṭāv upayanti. tany ekādaśa. sā triṣṭup.

yāni catvāri tany aṣṭāv upayanti. tāni dvādaśa. sā jagatī.

yāny aṣṭau sā gāyatrī.

This statement is a good example of both the operations of addition and subtraction. The examples can be multiplied; *cf.* for example, the RV. passage like 1.35.6 (*tisro dyāvā savituḥ, dvā upasthā, ekā yamasya bhuvane*) for 3-2 = 1 etc.

It must be noted, however, that there is neither any sign nor sign-word used to expressly indicate subtraction. It is by implication that we have to conjecture the process of subtraction. The only sign-word that is used for indicating the operation of subtraction is, as we have said before, *ūna*.

14

Examples of Multiplication

As we have said above, the process of multiplication is to be understood by the suffixes *-s* (i.e. *suc*), *-kṛtvaḥ* (i.e. *kṛtvasuc*) or the morpheme *-vṛt* or *-vāra* (both from *vṛ*) which are applied to the number words. No special word or sign is used to indicate the process. The following examples from all the nine Vedic saṁhitās may be cited for the working of the process of multiplication. The process can also be understood by implication, in which case the suffixes are not used at all.

14.1. Multiplication without suffixes

MS. 1.10.8 gives 12X3 = 36 and 12X2 = 24; *cf. ye vai trayaḥ saṁvatsarāḥ teṣām ṣaṭtriṁśat pūrṇamāsāḥ* (*saṁvatsara* = one year = 12; *cf.* MS. 3.3.3;4.6;7: *dvādaśa māsāḥ saṁvatsaraḥ*) *yau dvau tayoḥ caturviṁśatiḥ.*

TS. 7.4.11 = states 6X2 = 12; *cf. dvau ṣaḍahau bhavataḥ. tāni dvādaśa ahāni sampadyante* . also, 6X4 = 24; *cf.* TS. 7.4.11: *catvāraḥ ṣaḍahāḥ bhavanti. tāni caturviṁśati sampadyante* = KS.33.3. 10X2 = 20; *cf.* KS. 20.13: *yad viṁśati, dve virājau; daśākṣarā virāṭ.*

6X3 = 18; *cf. trayaḥ ṣaḍahā bhavanti; tany aṣṭādaśāhāni sampadyante,* KS. 33.3.

5X6 = 30; *cf.* KS. 33.3: *pañca ṣaḍahā bhavanti; tāni triṁśadahāni sampadyante.*

All these passages, as we have said earlier, go to show the method of counting by groups or sets. The phenomenon of 3X9 = 27 is expressed by a compound *triṇava* for 27 in KS. 33.8; *cf. saptaviṁśati grahītavyā*; triṇavā ime lokāḥ. This is an example in which a number is indicated by multiplication and not by addition, as in *sapta-viṁśati* (7+20). Many examples can be quoted. AV. 1.1.1 (*ye triṣaptāḥ pariyanti*) provides another example of a number indicated by multiplication; thus, 21 = 3X7.

14.2. Multiplication by the suffix -*suc* i.e. -*s*

20 expressed as 'two times ten' i.e. 2X10; *cf.* RV. 1.53.9: *tvam etān janarājñaḥ dviḥ daśa* .

21 expressed as 3X7; *cf.* RV. 1.72.6: *triḥ sapta guhyāni*.

10 as the multiplication of 2X5; *cf.* RV. 4.6.8;9.98.6: *dvir yat pañca svasāraḥ*. This phrase viz. *dviḥ paña* elsewhere occurs in the form of the result 10 of this multiplication; *cf.* RV. 9.1.7: *yoṣaṇo daśa*; RV. 3.29.13; 9.71.5 etc. *daśa svasāraḥ*.

The passage viz. RV. 8.96.8 (*triḥ ṣaṣṭis tvā maruto vāvṛdhānāḥ*) gives the multiplication of 3X60 = 180; an another passage, RV. 8.46.26 (*triḥ sapta saptatīnām*) states the product of ' three times seven seventies' i.e. (3X7)X70 = 1470; it can also be represented as 3X(70+70+70+70+70+70+70). A lot of examples of the multiplication by the suffix -*s* attached to the number-words are available from other *saṁhitās*, which need not be reproduced here, since the oldest *saṁhitā* viz. Ṛgveda is sufficient to prove the point; *cf.* MS. 4.2.4: *dviḥ ṣaṭ madhyataḥ* which gives 2X6 = 12; *cf.* also further MS. 4.2.4: *dvir daśa* = 2X10 = 20; KS. 38.11: *trir ekādaśa* - 3X11 = 33 etc.

For *catuḥ, 'four times'*, *cf.* AV. 11.2.9: *catur namo aṣṭakṛtvo bhavāya daśa kṛtvaḥ paśupate namaste, cf.* also *aṣṭakṛtvaḥ* (=8 times) and *daśakṛtvaḥ* (= 10 times).

14.3. Multiplication by the suffix *-kṛtvas*

The number-words to which the suffix *kṛtvas* (Pāṇinian *kṛtvasuc*) i.e. *kṛtvaḥ* is added to give out the sense of multiplication are: *tri, pañca, ṣaṭ, aṣṭa, nava, daśa, ekādaśa* and *dvādaśa; cf.* for *tri*, MS. 4.1.10: *triḥ kṛtvaḥ*; for *pañca*, TS. 6.1.1;9.5: *pañca kṛtaḥ*; for *ṣaṭ*, TS. 6.5.3: *ṣaṭ kṛtvaḥ*, for *aṣṭa*, TS. 6.4.5: *aṣṭau kṛtvaḥ*; for *nava*, MS. 4.5.7: *nava kṛtvaḥ;* for *daśa*, MS. 3.7.4: *daśa kṛtvaḥ;* for *ekādaśa*, TS. 6.4.5: *ekādaśa kṛtvaḥ*; and finally for *dvādaśa* also, TS. 6.4.5: *dvādaśa kṛtvaḥ*. In the case of the word *tri*, a peculiar fact notable is that both the suffixes viz. *s* and *kṛtvaḥ* are applied to it to make it a sort of double multiplicative. It is also to be noted that the multiplicative from *aṣṭa* with *kṛtvaḥ* is not *aṣṭakṛtvaḥ* but *aṣṭau kṛtvaḥ*, the nom./acc. suffix *au* remaining as it is. The suffix *kṛtvaḥ* is not found in the earlier three main *saṁhitās*, viz. RV. VS. and SV. AV.; AV. 11.2.9 (quoted above), however notes the form *aṣṭakṛtvaḥ* instead of *aṣṭaukṛtvaḥ*.

14.4. Multication by the suffix *-vṛt/vāram*

For the meaning 'once', *cf.* MS. 4.2.13: *tad ekavṛd aśayat*. The form, it is to be noted, is hapax in the whole of the Saṁhitā literature, and perhaps in the whole of the Sanskrit literature. As in the case of the suffix *-kṛt* where *eka* is substituted by *-sa*, so also in the case of the suffix *vṛt*, the word *eka* is substituted by *sa*; and we have the form *savṛt; cf.* KS.17.7: *savṛd asi; savṛte tvā.*

For *dvi-vṛt, cf.* KS. 11.4: *dvivṛt hiraṇyam dakṣiṇā; dvi-vṛt* = 'two times'.

For *tri-vṛt*, a host of references are available *cf.* RV. 1.140.2: *trivṛd annam ṛjyate* etc; *tri-vṛt*= 'three times'.

For *-vāra*, we have a word *śatavāra* in AV (19.36.1; 3; 6); but it refers to a *maṇi* i.e. jewel of that name. Yet, the repetition and context of the word *śata* many times in the hymn does not rule out the possibility of the word signifying 'a hundred times'.

14.5. Multiplication by the suffix *-kṛt*

This is available only in one case, that is, in the case of the word '*eka*' only. There also, the word *eka* is substituted by *sa* and we have the multiplicative as *sakṛt*. Many references throughout all the nine *saṁhitās* are available; *cf.* RV. 1.105.18: *aruṇo mā sakṛd vṛkaḥ pathā yantam dadarśa hi* , etc.

14.6. Other examples of multiplication

Besides the above examples in which the product of two numbers is indicated by some suffix, there are others in which the ready-made result of the multiplication is given without resorting to any suffixes suggesting multiplication. We have the examples from RV. and AV. in which we get the regular ready-made multiples of 2, 10 and 11. The following are the examples.

In RV. 2.18.4, we get regular multiples of the number 2, multiplied successively by numbers one to five; *cf.*

ā dvābhyāṁ haribhyām indra yāhi = 2 = 2X1;

ā caturbhiḥ = 4 = 2X2;

ā ṣaḍbhiḥ = 6 = 2X3;

ā aṣṭābhiḥ = 8 = 2X4; and

(ā) *daśabhiḥ* = 10 = 2X5

The next *ṛc.* viz. RV. 2.18.5 gives us all regular multiples of ten multiplied by numbers from one to ten; thus borrowing *daśabhiḥ* from the preceding *ṛc* quoted above, we have,

(ā) *daśabhiḥ* = 10 = 10X1;

ā viṁśatyā = 20 = 10X2;

(ā) *trimśatā* 30 = 10X3;

ā catvārimśatā = 40 = 10X4;

ā pañcāśatā = 50 = 10X5;

ā ṣaṣṭyā = 60 = 10X6;

(ā) *saptatyā* = 70 = 10X7;

(ṛc. 2.18.6) *ā aśītyā* = 80 = 10X8;

(ā) *navatyā* = 90 = 10X9 and finally

ā śatena = 100 = 10X10.

The AV. goes a step further and enumerates all the multiples of the number eleven in AV. 5.15.1-11. Thus we have,

ekā ca me daśa ca me - 1+10 = 1X11

dve ca me vimśatiś ca me = 2+20 = 22 = 2 (1+10) = 2X11

tisraś ca me trimsac ca me = 3+30 = 33 = 3 (1+10) = 3X11

catasraś ca me catvārimśac ca me = 4+40 = 44 = 4 (1+10) = 4X11

pañca ca me pañcāśac ca me = 5+50 = 55 = 5 (1+10) = 5X11

ṣaṭ ca me ṣaṣṭiś ca me = 6+60 = 66 = 6 (1+10) = 6X11

sapta ca me saptatiś ca me = 7+70 = 77 = 7 (1+10) = 7X11

aṣṭa ca me aśītiś ca me = 8+80 = 88 = 8 (1+10) = 8X11

nava ca me navatiś ca me = 9+90 = 99 = 9 (1+10) = 9X11

daśa ca me śatam ca me = 10+100 = 110 = 10 (1+10) = 10X11

Each of the second number in every line of the text is the one obtained by multiplying the first number by 10; thus in the first line, 10 is ten times of 1; in the second 20 is ten times 2; 30 is ten times 3 and so on. That this concept of 'ten times' is intended is clear by the fact that at the end of all these *ṛcs* comes the *pāda*, *śatam ca me sahasram ca me* , in which *sahasra* (= 1000) is ten times *śata* (= 100), and which totals to (1000+100 =) 1100 which is ten times of 110 given before.

This *ṛc* also reflects the idea of the process of multiplication of the first numbers mentioned in the *ṛcs* by numbers from 1 to 10. Thus, the numbers 1+10, taken as members of a single group are multiplied by numbers 1-10 independently or singly. Thus we have

1+10 = 1X(1+10) = 1+10 = 11

2+20 = 2X(1+10) = 2+20 = 22 and so on.

What seems to have been done is that, first, the number elevan is represented as the sum of 1+10. And then, at the second stage, the two numbers 1 and 10 indicating the parts of the number 11 are independently multiplied by numbers 1 to 10 and then lastly by 100. And we have the series–

1X (1+10) = 1+10 = 11

2X (1+10) = 2+20 = 22

3X (1+10) = 3+30 = 33 etc. and lastly ......

100X (1+10) = 100+1000 = 1100.

This certainly reflects a new approach or method of multiplication, because to say 2X11 = 22 is different from saying 2X(1+10) = 2+20 = 22. Such a method of representing the multiplication clearly reflects the concept of multiplying the parts of a number first and then totalling the multiplication results rather than multiplying the total. The parts which represent the parts of the big number are bracketed and each part is independently multiplied by a number outside the bracket.

From the other point of view, the number outside the bracket seems to have been taken out of the brackets as the common factor of the two numbers inside the brackets. Thus,

2+20 = 2 (1+10)

3+30 = 3 (1+10)

4+40 = 4 (1+10)

5+50 = 5 (1+10)

6+60 = 6 (1+10)

7+70 = 7 (1+10)

8+80 = 8 (1+10)

9+90 = 9 (1+10)

10+100 = 10 (1+10), and lastly

100+1000 = 100 (1+10)

Thus the factor (1+10) is common to all; as well as, the numbers 1-10 and 100 are common factors of the respectrive numbers. If this line of thinking is correct, we are led to the conclusion that the process of factorisation or taking out the common factor of the given expression/s seems to have been known to the Vedic people.[30] This method of representing the number (the number 11 in this particular context) gives us two ways of expressions: one can add any two numbers and then multiply them by a third number; or one can multiply by the third number the parts of a given number which can be suitably represented by analysing it into two (or three or even more) parts, and then add the results. Thus, to represent symbolially the principle by way of a formula, given the number *a* to be multiplied with the total of *c* and *d*, we can multiply first *a* independently with *c* & *d* and then add the results as $a \times c + a \times d = ac + ad$, or multiply the total of *c* & *d* as $a \times (c+d) = ac+ad$, which again will be equal to the above result.

Also, the number 11, for example, is a substitute for the sum 1+10, or vice versa, the expression 1+10 can be taken as the substitute of 11. In Pāṇinian terminology, 1+10 is an *ādeśa* of 11, and conversely, 11 is an *ādśa* for 1+10[4].

The same process noted above finds its exact, but inverse or upside-down, replica or mirror-image in the AV. 19.47.3. What is done in AV.19.47.3 is that the author has started from the number 99 and come down to the number 11 which is the lowest (*avama*). The *ṛc* is as follows:

ye te...draṣṭāro navatir nava = 90+9 = 99

aśītiḥ...aṣṭa = 80+8 = 88

sapta saptatiḥ = 7+70 = 77

ṣaṣṭiś ca ṣaṭ ca = 60+6 = 66

pañcāśat pañca = 50+5 = 55

catvāraś ca catvāriṁśac ca = 4+40 = 44

trayas triṁśac ca = 3+30 = 33

dvau ca viṁśatiś ca = 2+20 = 22

te...ekādaśa avamāḥ = 1+10 = 11

In this case, the author has started from the highest single-dight multiplier viz. 9 and came down gradually to the lowest single-digit multiplier or common factor 1; and we have this descending series of the multiples of 11.

The passage from VS. 18.25 states the multiples of the number 4. It runs as follows:

catasraś ca me, aṣtau ca me,...dvādaśa ca me,...ṣodaśa ca me,...viṁśatiś ca me,...caturviṁśatiś ca me,...aṣṭāviṁśatiś ca me,...dvātriṁśat ca me,...ṣaṭtriṁśat ca me,...catvāriṁśat ca me,...catuścatvāriṁśat ca me,...aṣṭācatvārīṁśat ca me.....

Thus, we have 4, 8, 12, 16, 20, 24, 28, 32, 36, 40, 44 and 48 as the multiples of 4, which are obtained by multiplying 4 by numbers from 1 to 12.

It is really interesting to note how the multiples of 4 are derived by Veda. The root of the derivation of these multiples lies in the previous verse, viz. VS. 18.24. It lists all the odd numbers from 1 to 33 and reads as follows:

ekā ca me, tisraś ca me...pañca ca me...sapta ca me...nava ca me... ekādaśa ca me... trayodaśa ca me... pañcadaśa ca me... saptadaśa ca me... navadaśa ca me... ekaviṁśatiś ca me... trayoviṁśatiś ca me,... pañcaviṁśatiś ca me... saptaviṁśatiś ca me... navaviṁśatiś ca me... ekatriṁśat ca me... trayastriṁśat ca me....

We state the above two verses of VS. in modern symbols for numbers for the sake of convenience and easy understanding.

VS.18.24: 1, 3, 5, 7, 9, 11, 13, 15, 17, 19, 21, 23, 25, 27, 29, 31, and 33.

VS. 18.25: 4, 8, 12, 16, 20, 24, 28, 32, 36, 40, 44 and 48.

If VS. 18.25 is read with and in the context of the previous verse, VS. 18.24, one can easily find out the roots and the method of the derivation of the multiples of 4.

Looking closely and comparing the two verses 18.24, and 25, we find that the first number 4 of the *ṛc* 18.25 is derivable by the addition of the first two odd numbers viz. 1 and 3 of the verse VS. 18.24. The next second multiple of 4 viz. 8 can be derived by the addition of the next odd numbers 3 and 5; the third multiple by adding the next odds viz. 5 and 7; the fourth by the addition of 7 and 9 and so on. It is because mathematically the verse 18.25 depends upon the verse 18.24 for a clear meaning that the two verses are put together, 18.25 succeeding 18.24.

The MS. 1.5.8 has an interesting story to tell about both the operation of addition as well as multiplication. To quote,

manor ha vai daśa jāyāḥ āsan: daśaputrā, navaputrā, aṣṭaputrā, saptaputra, ṣaṭputrā, pañcaputrā, catuṣputrā, triputrā, dviputrā, ekaputrā. Ye nava āsan tān ekaḥ *upasamakrāmat; ye aṣṭau tān dvau; ye sapta tān trayaḥ; ye ṣaṭ tān catvāraḥ; atha vai pañca eva pañca āsan; tāḥ imāḥ pañca, daśataḥ imān paṇca nirabhajan; yad eva kiṁ ca manoḥ svam āsīt, tasmāt te vai manum eva upādhāvan; manau anāthanta. tebhyaḥ etāḥ samidhaḥ prāyacchat. tābhir vai te tān niradahan. tābhir enān parābhāvayan.*

Freely translated, what the story tells is: Manu had ten wives, named respectively as *dasaputrā* (having ten sons), *navaputrā* (having nine sons), *aṣṭaputrā* (having eight sons), *saptaputrā* (having seven sons), *ṣaṭputrā* (having six sons), *pañcaputrā* (having five sons), *catuṣputrā* (having four sons), *triputrā* (having three sons), *dviputrā* (having two sons) and *ekaputrā* (having one son).

The one and the only son of *ekaputrā* merged with nine of the *navaputrā*; the two sons of *dviputrā* joined with the eight of *aṣṭaputrā*; the three sons of *triputrā* combined with the seven of the *saptaputrā*; the four sons of *catuṣputrā* crossed over to the six of the *ṣaṭputrā*. The five sons of *pañcaputrā* were left alone, against all others.

The five sons of the *pañcaputrā* then approached *Manu*. Manu had some property/quality (*cf.* the word *svam* which conveys the sense of the inherent natural quality) of his own. He gave the five sons the *samidhs* i.e. the oblation-sticks. They took the *samidhs* and defeated all the other sons.

Here the story ends. The whole story, it should be noted, is a kind of commentory or explanation of the Vedic *ṛc*, *indhānās tvā śatam himāḥ dyumantam samidhīmahi* , which occurs first in VS. 3.18 and is repeated in TS. 1.5.5,7; MS. 1.5.2,8; and KS. 6-9, 7.6, 35.2. AV 19.55.4 reads: *indhānās tvā śatahimā ṛdhema* .

Representing the number-words in the above story in modern number-symbols, the picture we get is as follows:

10 wives: 10, 9, 8, 7, 6, 5, 4, 3, 2, 1

The number 1 joining with 9 gives us 10

The number 2 joining with 8 gives us 10

The number 3 joining with 7 gives us 10

The number 4 joining with 6 gives us 10

The left-hand side of the numbers from 10 to 6 thus gives the numbers as 10, 10, 10, 10, 10 which all make together 50. There is only one number left over on the right-hand side, and that is 5, which, being isolated, becomes panicky and approaches Manu for guidance and protection as there are 50 on the other left-hand side against 5 on the right-hand side. In other words, the strength of the left-hand side is ten times that of the right-hand side. Manu gives the number 5 a trick, a device, a remedy and the remedy is in the form of the *samidhs*.

What is this *samidh*? The *samidh* in this context refers to the *ṛc* VS 3.18 quoted above; this *ṛc* contains the word *śatam* i.e. 100. Manu gives the number 5 this trick to become hundred-fold as against the ten-fold of the other side. The number 5 then becomes hundred-fold into 500 and wins over a defeat on the number 50.

The story clearly states two mathematical operations, viz. addition of 9, 8, 7 and 6 with respectively 1, 2, 3 and 4 giving out 10 in each case; and multiplication or repetition of the number 5 hundred times giving out the number 500.

We can thus see how the scientific truths stated in the Vedas are shrouded under the garb of short stories which apparently look mythical and childish, but which help us in arriving at the core of the theories they aim at expounding. The Vedic verses contain some suggestive word/s which help us in their interpretation. The context and the tone of the stories is very important in such case. The present story of Manu, his wives and sons, for example, contains and is playing with numbers. We, therefore, select the number-word *śatam* from the verse which Manu recommends to the 5 sons. If, however, the context of the story were pertaining to seasons, the word from the given verse important and suggestive for us would have been *himāḥ* which means 'cold autumns'. But since the context and contents of the story are numbers, we choose the number-word *śatam* to get the clue.

The story can also be interpreted in another way by basing the interpretation on the mathematical operation of addition. We borrow the word *śatam*, and consequently the number 100 also from the given *ṛc* and add it to the number 5 which is left over alone after all the other number joined together to make 50. And we have 5+100 = 105, which is more than double the number 50 on the left-hand side. The number 105 then defeated the number 50; that is to say, 105 outnumbered 50.

In view of the fact of addition of all other numbers, the present interpretation based on the addition of 5 with 100 is also not unwarranted. Since the other numbers swelled themselves by addition, Manu advised number 5 to join with or add with the number 100.

15

The Examples of Division

As we have seen before, the suffix which is employed by the Veda to suggest or signify the process of division is *-dhā* which is applied to the number-words; and we have the divisional distributives as *ekadhā, dvidhā* etc. The following examples provide us with the evidence of the knowledge of the mathematical procedure of division on the part of the Vedic sages.

15.1. *ekadhā*

AV.5.17.8 (*uta yat patayo daśa striyāḥ pūrve abrāhmaṇāḥ, brahmā ced hastam agrahīt sa eva patir ekadhā*) says that a Brahmin is the only one husband for a lady. Literally, *ekadhā* would signify 'in only one way'; *cf.* also AV. 8.9.26 (*ekam dhāma, ekadhā āśiṣaḥ*) and AV. 10.10.5 (*ye devās tasyām prāṇanti te vaśam vidur ekadhā*). The meaning of *ekadhā* seems to fluctuate between 'in one way' and 'one time'; *cf.* also MS. 4.3.8: *ekadhā vā etad yajamāne yajñasyāśīḥ pratitiṣṭhati* .

15.2. *dvidhā/dvedhā/dvaidha*

The distributive from *dvi* occurs in non-numerical context in all the passages. Mathematically speaking, we have no example for *dvidhā*; *cf.* RV. 10.56.6. and also other *saṁhitās.*

Yet, the meaning conveyed by *dvidhā*, viz. 'in two parts' is conveyed by the word *ardha* 'half'. Perhaps because of the existence of the word *ardha* the Vedic people did not find it necessary to use the word *dvidhā*, a derivative from *dvi*. Incidentally, the word *ardha* is derived from **ṛdh* 'to prosper, to grow' and with a semantic change 'to become many'. In dividing a thing/sum into two parts, one makes it into many.

The examples *dviḥ pañca* (=10) and *dvau ṣaḍahau* (= 12), quoted previously in the context of the operation of multiplication can be cited in the context of the operation of division also, since division is the reverse process of multiplication. Thus, 5x2 = 10 and 10/2 = 5 or 10/5 = 2; 6x2 = 12 and 12/2 = 6 or 12/6 = 2 etc.

For *ardha* as 'half', which also occurs always in the non-numerical context, *cf.* RV. 1.92.1, 2.30.5, 6.30.1 etc. RV. 1.92.1 mentions 'two halves' and refers to one half-part as '*pūrva ardha*'; RV. 1.164.12 refers to the second half-part as '*para ardha*'. RV. 2.27.15 refers to both the parts as *ubhau ardhau*. MS. 4.6.6 says that Indra became into two parts; *cf. iti sa dvibhāgaṁ babhūva* . He is, therefore, called *ardhabhāk cf.* MS. 3.4.1.

The following passage from TS. 7.1.5 is very interesting from the point of view of the division of the number 1000 into three parts. The numerical value of the parts, however, is not given there; it simply refers to the division of 1000 into three parts. The passage runs as follows:

atha yā sahasratamī āsīt tasyām indraś ca viṣṇuś ca vyāyachhetām/sa indro amanyata: idam viṣṇuḥ sahasram varkṣyate iti. tasyām *akalpetām dvibhāge indraḥ , tṛtīye viṣṇuḥ,* which, roughly rendered, is as follows:

Both Indra and Viṣṇu laid their claim on (something) which was one thousand. Indra thought that Viṣṇu would snatch away (*varkṣyate,* from √*vṛc/vr*) the whole one thousand. So they decided–two parts for Indra and the third part for Viṣṇu.

The same passage further says that 2/3 should be given to brahman and 1/3 to the Agnīdh; *cf. dvibhāgam brahmaṇe, tṛtīyam agnīdhe.* How the number 1000 is divided into 3 equal parts, except in fraction, is not clear. The 1/3rd part will only be 1000/3. MS. 4.6.6 (*eṣa ardhabhāk prāṇānām*) mentions the fraction 1/2 by the word *ardhabhāk*, 'divided into half'. Now, the interesting point is that the *prāṇas* are said to be *daśa* i.e. ten in the immediately preceding line (*daśa vai prāṇāḥ*) The 'half' of the *prāṇas,* therefore, comes to 5.

In the same passage of MS. 4.6.6, we get the following references to the 1/3 and 2/3 parts of an entity. Not only this, but the last line says that they together become full. *cf. divaḥ āpyāyasva iti sa tṛtīyam babhūva. antarikṣād āpyāyasva iti sa dvibhāgam babhūva. pṛthivyā āpyāyasva iti sa pupūre,* which can be roughly rendered as:

(The Adhvaryu said) "You swell from the heaven"–and he became one-third. (Adhvaryu said) "You swell from the mid-region"–and he became two-parts. (Adhvaryu said) "You swell from the earth"–and he became full or whole.

From the two passages of TS and MS. quoted in full it is very clear that the *ṛṣis* knew the fractions 1/3, 2/3 and 1/2. It is also/ evident that they knew full well that 1/3+2/3 together make one whole i.e. 1. TS. 7.4.1 (= KKS.31.20) mentions that a *saṁvatsara* i.e. an year (=12 months) consists of 24 half-months. This gives a division of 24 by 2 as 24/2 = 12 which is the number of months of an year.

15.3. *tridhā/tredhā/traidha*

These distributive adverbs nowhere occur in the numerical context in any of the Vedic *saṁhitās*. The passages quoted in the case of the multiplication-process can also be cited for the process of division. Thus, *triḥ sapta* = 3X7 = 21; and 21/3 = 7 or 21/7 = 3 etc.

15.4. *caturdhā*

The *Ṛbhus* in the RV. are said to divide one cup (*camasa*) into four parts; *cf.* RV. 4.35.2: *ekam vicakra camasam caturdhā* ; also RV. 4.35.3: *vyakṛṇota camasaṁ caturdhā* . The example *catvāraḥ ṣaḍahāḥ* (= 24), quoted in the context of multiplication can also be borrowed here to show division; thus 4X6 = 24; therefore, 24/4 = 6 or 24/6 = 4.

15.5. *pañcadhā*

The passage *pañca ṣaḍahāḥ bhavanti* quoted before can be cited for the operation of division also; thus, 5X6 = 30; hence, 30/5 = 6 or 30/6 = 5.

15.6. *division by seven*

The Maruts, whose number is given by RV. (1.133.6: *trisaptaiḥ śūra satvabhiḥ*) as 21 always stay in groups. The total number of groups of the Maruts, as TS. 2.2.11 (*saptagaṇāḥ vai murutaḥ...gaṇaśaḥ eva etān avarundhe*), is 7. Each group, therefore, automatically consists of 3 Maruts. The number 21 here is seen to be divided by 7, and we have, 21/7 = 3. RV. 5.52.17, 8.28.5, however, (*cf. A.A. Macdonell, Vedic Mythology* , p.78) gives the number of Maruts as 'thrice sixty' (*cf.* RV. 8.85.8). RV. 5.52.17 and 8.28.5 give the number of Maruts as 49 (*sapta-sapta*) *cf.* also *Devībhāgavata* , 4.3; *Matsya Purāṇa*, 7 and *Rāmāyaṇa*, 1.45,46; *cf.* also *SK Lal, Female Divinities in the Hindu Mythology and Ritual*, CASS, 1980, p. 14. In that case, the number 49, grouped into 7 groups, will be divided by 7; thus, 49/7 = 7. The word *gaṇa* shows the idea of forming or counting by groups.

We have other distributive words also like *navadhā, daśadhā, sahasradhā* etc. But we do not find any convincing evidence, especially in numerical contexts, which will unmistakably show the operation of division. Yet, since the words are used, they indicate the knowledge of division of a number in so many parts as the number to which the suffix *-dhā* is added indicates.

15.7. Division into many parts

A passage from TS. 5.4.8 (*bhūyiṣṭhabhāktam indram dadhāti*) uses the word *bhūyiṣṭha* which is the superlative of the word *bahu* and/or *bhūri*. The word means 'Indra who is divided into many'. This word indicates the division of an entity into many parts. The passage mentioning Indra's division into two parts may be recalled here for his division into many parts.

16

The Concept of Fractions

Just as the two processes of multiplication and division, like the two processes of addition and subtraction, are inverse reflections of each other, and one implies the other, the process of division also implies the knowledge of yet another mathematical category; and the category or the concept is 'the fraction'. Actually knowledge of fractions is a necessary, inevitable corollary of the knowledge of division. The words which signify the 'fraction' used in later Sanskrit mathematical literature are '*aṁśa*' (i.e. part) and *mātrā* (lit. measure; but it also means 'part'). Bhāskarācārya uses the word '*bhāgajāti*', 'the parts'. (*cf. Leelavatī*, verse 30). The word *mātrā* in the above-mentioned sense is used in the Vedic literature. *Cf.* TS. 7.1.6. *tredhāvibhaktam vai trirātre sahasram sāhasrīm eva enāṁ karoti; sahasrasya eva enāṁ mātrām karoti* , which signifies the 1/1000th part. In RV. the word *mātrā* signifies just 'a measure', whose numerical value is unknown (*cf.* RV. esp. 3.38.3; 46.3; 10.71.11). Yet the meaning of *mātrā* as 'part' (= *aṁśa*) is quite evident from the passages of other *saṁhitās* like the TS. quoted above. Other references from other *saṁhitās* may also be quoted. In MS. 4.6.5 (*tryanīkam asya prajā bhaviṣyati*), the word *tryanīka* signifies 'three parts'.

Another peculiarity of Vedic words for indicating fraction is that they use the same words for showing 'that much part' of the number, as are used for ordinals of the number-words. Thus, in TS. 2.4.12, we have the word *tṛtīya* itself, meaning not 'the third' as ordinal, but 'the third part'; *cf. sa viṣṇuḥ tredhā ātmānam vi nyadhatta. pṛthivyām tṛtīyam* (i.e. 1/3 on the earth), *antarikṣe tṛtīyam* (i.e. 1/3 in the mid-region), *divi tṛtīyam* (i.e. 1/3 in the heaven). It is clear that the word *tritīya* is used to signify 'the third part'. Also, the words for indicating 'the first, second or third part', are not used at all. The first *tṛtīya*, therefore, means 'the first 1/3rd;' the second *tṛtīya* means 'the second 1/3rd' and the last *tṛtīya* means 'the last 1/3rd.' The word *tṛtīya*, therefore, serves the function of what we call 'the denominator' in modern mathematics; and the word for numerator is taken to be understood from the contexts. That is to say, in the first *tṛtīya*, the numerator is 1; in the second it is 1 and in the third also, it is 1.

We have many examples in the *saṁhitās*. The above example of 1/1000th part has a paralled in RV. 6.69.8 (*tredhā sahasram...airayethām*) *cf.* also TS. 5.2.6: *vṛtraḥ...tredhā abhavat. sphyaḥ tṛtīyam, rathas tṛtīyam, yūpaḥ tṛtīyam,* which is identical with TS 6.1.3; KS. 20.4; KKS. 31.6. The same meaning of 1/3 is conveyed by the word *tṛtīya* by MS.4.6.2: *tad vai bheṣajam tredhā vi nyadadhuḥ. agnau tṛtīyam, brāhmaṇe tṛtīyam, apsu tṛtīyam. cf.* also, KS. 23.6: *tredhā vā etasya pāpmānam vibhajante...yo annam atti sa tṛtīyam, yo aślīlam kīrtayati sa tṛtīyam, yo nāma gṛhṇāti sa tṛtīyam. cf.* also KKS 7.3: *tredhā tanvo vi nyadhatta. paśuṣu tṛtīyam, apsu tṛtīyam, āditye tṛtīyam;* cf. also KKS. 35.8: *yajñaś ca...tredhā prāviśat. ṛcam tṛtīyena, sāma tṛtīyena, yajus tṛtīyena.*

The fraction 1/2 is indicated by the word *ardha* 'half'. The word is found with two accents–one as *ardhá* accenting on the last syllable, and the other as *árdha* accenting on the first syllable. There is yet no difference of meaning between the two words. Two halves make 'one full;' *cf.* AV. 5.1.9: *ardham ardhena payasā pṛṇakṣi, ardhena śuṣma vardhase asura, cf.* also AV. 10.8.7: *ekacakram vartate ekanemi...ardhena viśvaṁ bhuvanam jajāna, yad asya ardham kva tad babhūva;* also AV. 10.8.13: *ardhena viśvam bhuvanam jajāna, yad asya ardham katamaḥ sa ketuh.*

We do not get explicit reference to other simple fractions like 1/4, 1/5, 1/6 etc. The following passages from AV. may, however, be cited; *cf.* AV. 15.15.1: *tasya vrātyasya. sapta prāṇāḥ yo asya prathamaḥ prāṇaḥ* (= first of the seven = 1/7); *yo asya dvitīyaḥ prāṇaḥ* (= 2nd of the seven = 2/7) etc. The list goes upto *saptama* (*cf.*AV. 15.15.9) which gives us 7th of the seven, i.e. 7/7 which is equal to unity. The same idea is repeated in the hymns AV. 15.16 and 15.17. We, therefore, may get the fractions 1/7, 2/7, 3/7, 4/7, 5/7, 6/7 and 7/7i.e. unity.

AV. 11.1.5 (*tredhā bhāgo nihitaḥ yaḥ purā...aṁśān jānīdhvam, vi bhajāmi*) states that "three parts were laid down, in ancient times,...I divide them; know the parts." We have seen before (in the case of the word *tṛtīya* and *tredhā*) that while denoting the fraction 1/3, the word for the denominator (viz. *tṛtīya*) is used. But sometimes the word for the numerator also is used; the denominator, being common, is left to be understood. Thus, in AV. 5.2.8.6 (*tredhā jātam janmanā idam hiraṇyam...agner ekam priyatamam babhūva, somasya ekam, apām ekam*); it is said that *hiraṇya* was born in three parts; one (part) became dear to Agni, one to the Soma and one to the waters. Thus instead of speaking *tṛtīyam*, the *ṛc* says *ekam* i.e. `one' referring to the numerator in the fractions, 1/3, 1/3, 1/3. It is also clear from all the above passages that the Vedic people knew that 1/3+1/3+1/3 give out 3/3 i.e.1 or unity.

Besides indicating the division by the numerator number-word or denominator number-word, we have statements in the form of phrases which bring out the idea of parts. Thus, in RV. 10.94.4 (*pādo'sya viśvā bhūtāni tripād asyāmṛtam divi*) we have the fractions 1/4 (*pādaḥ*) and 3/4 (*tri-pād*). RV. 10.27.16 gives us the fraction 1/10; *cf. daśānām ekam* (= one out of ten). RV. 3.2.9 gives 1/3; *cf. tisro yahvasya samidhaḥ tāsām ekām adadhuḥ* ; also RV. 10.5.6: *sapta maryādāḥ...tāsām ekam* . The idea of 'one' divided into 'many' is conveyed by RV. 10.114.5: *ekam santam bahudhā kalpayanti*. RV. 4.35 and 4.36 elaborate the idea of 'one *camasa*' divided into 'four' (*caturdhā* or *caturvayam*) parts. The word *purudhā* (= divided into many or many ways) is also used as a synonym of *bahudhā*.

In non-numerical context, we see the word *bhāga* and *bhaj* is mostly used; *cf.* RV. 3.49.4: *vibhaktā bhāgam*; RV. 1.123.3: *bhāgam vibhajāsi*. RV. 5.44.12 states the division of a group (*gaṇa*); *cf. gaṇam bhajate.* Agni and Savitṛ are said to be the *vibhaktṛ* i.e. 'one who divides, distributes, separates or analyses' and we have phrases like *dhanaṁ vibhaktā, rāyo vibhaktā, vasvo vibhaktā, ratnam vibhaktā* in the Ṛgveda, which all mean 'distributor of wealth'. We have also the idea of 'distributing' the abstract things like *vāja* (strength, RV. 6.36.1 etc.), *śravas* (fame, RV. 7.18.24 etc.) and *śramasya dāya* (the part of labour i.e. remuneration(?), RV. 10.114.10).

We can see from all the above references to the process of division and to fractions, that the Vedic people knew these operations. They also knew, as we have seen before, the forming of groups or sets and counting them again.

17

Squares, Square-roots, Cubes and Cube-roots

So far as the reference to the concepts and processes of the mathematical operations of squares and square-roots, and cubes and cube-roots are concerned, we do not get any direct reference to them, nor do we get any technical term used to indicate them. There are absolutely no references to the operation of square-roots, cubes and cube-roots. The only reference pertaining to the formation of squares is again indirect and is not from the numerical context; it is from ritual context. Though it is so, interpreted from the point of view of forming squares, it gives us an easy method of forming squares. There are two references to the same numbers; and they are VS. 14.28-31 and VS. 18.24. They read as follows:

VS.14 28-31:

ekayā astuvata... tisṛbhiḥ... pañcabhiḥ... saptabhiḥ... (28)... *navabhiḥ... ekādaśabhiḥ ... trayoda-śabhiḥ... pañcadaśabhiḥ... saptadaśabhiḥ* (29)... *navadaśabhiḥ ... ekaviṁśatyā... trayo-viṁśatyā... pañcaviṁśatyā... saptaviṁśatyā* (30)... *nava-viṁśatyā... ekatriṁśatā trayastriṁśatā* (31).

VS. 18.24.

ekā ca me tisraś ca me, tiśraś ca me pañca ca me, pañca ca me sapta ca me, sapta ca me nava ca me, nava ca me ekādaśa ca me, ekādaśa ca me trayodaśa ca me, trayodaśa ca me pañcadaśa ca me, pañcadaśa ca me saptadaśa ca me, saptadaśa ca me navadaśa ca me, navadaśa ca me ekaviṁśatiś ca me, ekaviṁśatiś ca me trayoviṁśatiś ca me , trayoviṁśatiś ca me pañcaviṁśatiś ca me, pañcavīṁśatiś ca me saptaviṁśatiś ca me, saptaviṁśatiś ca me navaviṁśatiś ca me, navaviṁśatiś ca me ekatriṁśat ca me, ekatriṁśat ca me trayastriṁśat ca me...

If we compare both the above passages, we find a striking similarity. Barring the differences of the case-endings of the number-words, numerically speaking, both the passages are not only similar but identical. Both of them note the odd numbers only. Both of them note the same numbers. And both of them start from number 1 and go only upto the odd number 33. What do these passages drive us to conclude mathematically? To get the answer, we write the number-words in modern number-symbols. The verses contain the following odd numbers:

1, 3, 5, 7, 9, 11, 13, 15, 17, 19, 21, 23, 25, 27, 29, 31 and 33.

Out of the two passages, the passage VS. 18.24 helps us to read between the lines mathematically. What the passage VS. 18.24 contains is especially the word *ca* 'and', which as we have seen before, points to the mathematical operation of addition. Moreover, the passage VS. 18.24 contains the repetition of the odd numbers 3 to 31. Combining these two clues, we go on adding the numbers–first, the first two numbers, then the first three numbers and so on. We write the additions separately to bring out the conclusions clearly. The picture that emerges is as follows:

(1+0 = 1)

1+3 = 4

1+3+5 = 9

1+3+5+7 = 16

1+3+5+7+9 = 25

1+3+5+7+9+11 = 36

1+3+5+7+9+11+13 = 49

1+3+5+7+9+11+13+15 = 64

1+3+5+7+9+11+13+15+17 = 81

1+3+5+7+9+11+13+15+17+19 = 100 and so on upto 1+3+...+33 which gives us the total as 289, which is = 17^2; 17 is the number of odd terms from 1 to 33.

If now we look at the right-hand-side results of the additions, we find that they are all squares of the numbers from 2 to 17. We do not for certain know what the *ṛc* is aiming at or implies; yet, read mathematically, the verse certainly gives out the procedure for finding out the squares of the numbers from 1 onwards.

The general rule which can be spelled out for finding the squares of the numbers may be stated as follows:

Add the odd numbers beginning from 1 in groups of two, three, four etc. and we get the squares of the number of odd numbers added together. Thus, if we add the first two odd numbers viz. 1 and 3, we get the square of two; if we add the first three odd numbers, viz. 1, 3 and 5, we get the square of the number three and so on.

Symbolically, if $(x)_{n1}$, $(x+2)_{n2}$, $(x+4)_{n3}$...is the series, in which x, x+2, x+4 etc. are the odd numbers and n indicates the number of the terms, the general formula for finding out the squares of the numbers beginning from 1 can be spelled out as:

$$(x)_{n1}+(x+2)_{n2}+(x+4)_{n3}\ldots+(x+4n)_{nn} = n^2$$

Such may be process laid down by VS. 14.28 — 31 and VS. 18.24 for finding out the squares of the numbers from 1 onwards.[31]

Besides this, we do not get any reference to the concept or procedure of finding out squares of positive integers.

The square-roots, cubes and cube-roots are absolutely nowhere mentioned, either explicitly or implicitly, in any of the nine Vedic Saṁhitās taken here for study.

18

Arithmetic and Geometric Progression

The Veda states certain series which in modern mathematical terminology can be termed as the 'arithmetic progression' and 'the geometric progression'.

18.1 Arithmetic progression

Arithmetic progression is defined as " sequence of numbers in which each number is larger (or smaller) than the number that precedes it by a constant amount. The increase (or decrease) is called the *common difference.* Examples: 1, 3, 5, 7, 9...(in which the common difference is 2); 25, 22, 19, 16...(in which the common difference is -3). If the terms of the progression are in an increasing order, the common difference is *positive*; if decreasing, *negative.* The last term of an arithmetic progression is given by the formula

$$l = a + (n-1)\, d,$$

where l = last term, a = first term, n = number of term in the series and d = common difference...

The first known discussion of arithmetic progression occurs in the Egyptian *Ahmes Papyrus* (c. 1550 B.C.)....A rule for finding the sum of an arithmetic series was developed around 510 AD. by the Hindu mathematician Aryabhata (the Elder)".[32]

It may, however, be noted that the Egyptian *Ahmes Papyrus* is not the first to mention and discuss the arithmetic progression.

The passages from RV. VS. and AV. quoted before may be cited as the best examples of the concept of arithmetic progression in Vedic times.

RV. 2.18.4 and 2.18.5 give us the arithmatic progression in the enumeration from, first 2 to 10 and then from 10 to 100. We have thus the following two series of arithmetic progression from the RV.:

2, 4, 6, 8 and 10...I, and then

10, 20, 30, 40, 50, 60, 70, 80, 90 and 100...II.

The difference between any two consecutive number-members in series I is 2 and in series II is 10.

AV. 5.10.1-11 gives us the following series of arithmetic progression:

11, 22, 33, 44, 55, 66, 77, 88, 99 and 110.

The difference between any two consecutive numbers in the series is constant i.e. 11.

VS. 18.24 gives us the following series of arithmetic progression:

1, 3, 5, 7, 9, 11, 13, 15, 17, 19, 21, 23, 25, 27, 29, 31 and 33.

The difference between any two consecutive numbers is 2.

VS. 18.25 states the following series of arithmetic progression:

4, 8, 12, 16, 20, 24, 28, 32, 36, 40, 44 and 48.

The difference between any two consecutive numbers is 4.

All these series are in the ascending order, the succeeding number being greater than the preceding one.

AV. 19.47.3 gives the following series:

99, 88, 77, 66, 55, 44, 33, 22 and 11.

The difference between any two consecutive numbers is constant i.e. -11. This series is in the descending order, the succeeding number being less than the preceding one.

It should also be borne in mind that though the ascending series given here are stated to a finite limit, being given the difference between two consecutive numbers, they have the potentiality of being expanded to infinity. There can thus be infinite number of infinite series of arithmetic progression. Such potentiality also exists in the descending series also.

The series of natural numbers viz. 1, 2, 3, 4...etc., as we have said earlier is also in arithmetic progression, with the difference of 1 between any two consecutive members.

All the above series of arithmetic progression give us the general formula–

x, x+d, x+2d, x+3d...+x+nd...*ad infinitum*.

The TS. 7.2.11-19 provides many examples of arithmetic progression. They are given in the following:

Examples from TS.:

The following passages from TS. 7.2.11-20 are especially notable for the arithmetic progression. The passages are quoted in full for ready reference:

TS.7.2.11: *ekasmai svāha, dvābhyām svāhā, tribhyaḥ svāhā, caturbhyaḥ svāhā, pañcabhyaḥ svāhā, ṣaḍbhyaḥ svāhā, saptabhyaḥ svāha, aṣṭābhyaḥ svāha, navabhyaḥ svāhā, daśabhyaḥ svāhā, ekādaśabhyaḥ svāhā, dvādaśabhyaḥ svāhā, trayodaśabhyaḥ svāhā, caturdaśabhyaḥ svāhā, pañcadaśabhyaḥ svāhā, ṣoḍaśabhyaḥ svāhā, saptadaśabhyaḥ svāhā, aṣṭādaśabhyaḥ svāhā, ekānnaviṁśatyai svāhā,...*

Upto here i.e. 19, for which TS uses the word *ekānnaviṁśati* (i.e. 20-1), the numbers are serially arranged; then suddenly TS jumps to the numbers 29, 39, 49, 59, 69, 79, 89 and 99, because once the method of building up the series is given, the rest of the numbers are to be automatically built up and named on those lines. To explain, once the numbers from 11 to 19 are to be named as *ekādaśa, dvādaśa* etc., it is needless to lay down explicitly the numbers from 20 onwards as *eka-viṁśati, dvāvīṁśati* etc., simply because these later numbers are also to be formed on the pattern of addition of *eka, dvā* etc. with *vīṁśati*. Hence the jump from 19 to 29 in the following:

navavīṁśatyai svāhā, ekānnacatvāriṁśate svāhā, navacatvāriṁśate svāhā, ekānnaṣaṣṭyai svāhā, navaṣaṣṭyai svāhā, ekānnāśītyai svāhā, navāśītyai svāhā, ekānnaśatāya svāhā, śatāya svāhā,... then suddenly the jump is to 200 as *dvābhyām śatāya svāhā*. It is to noted that the ninth number in each series is alternately given with *ekānna-* and *nava-*; thus,

19 = *ekānnaviṁśati* (=20-1, back-counting)

29 = *navaviṁśati* (20+9, forward counting)

39 = *ekānnacatvāriṁśat* (40-1, backward counting)

49 = *navacatvāriṁśat* (49+9, forward counting)

59 = *ekānnaṣaṣṭi* (60-1, backward counting)

69 = *navaṣaṣṭi* (60+9, forward counting)

79 = *ekānnāśīti* (80-1, backward counting)

89 = *navāśīti* (89+9, forward counting)

99 = *ekānnaśata* (100-1, backward counting)

The purpose behind alternatively mentioning the names of the ninth number in each series of tens seems to be to acquaint the students with both the methods of counting viz. forward counting and backward counting.

The sudden leap from 100 to 200 seems to suggest that the numbers between 100 to 200 are to be formed according to pattern already stated in the previous lines. This is an instance of arithmetic progression.

TS.7.2.12: *ekasmai svāhā, tribhyaḥ..., pancabhyaḥ..., saptabhyaḥ..., navabhyaḥ..., ekādaśabhyaḥ..., trayodaśabhyaḥ..., pañcadaśabhyaḥ..., saptadaśabhyaḥ..., ekānnaviṁśatyai...*

From here the jump is to 29.

navaviṁśatyai... ekānnacatvāriṁśate... navacatvāriṁśate..., ekānnaṣaṣṭyai..., navaṣaṣṭyai..., ekānnāśītyai..., navāśītyai..., ekānnaśatāya... śatāya svāha.

The alternate backward and forward counting is notable; the numbers 1, 3, 5..., which are all odd numbers,[33] represent, as we have seen before, a series of arithmetic progression, with a difference of +2 between any two consecutive numbers. Besides as in the case of the passage from VS. 18.24, they help us to find out the squares of the numbers from 1 to any limit.

TS.7.2.13: notes the even numbers:

dvābhyām svāhā..., caturbhyaḥ..., ṣaḍbhyaḥ..., aṣṭābhyaḥ..., daśabhyaḥ..., dvādaśabhyaḥ..., caturdaśabhyaḥ..., ṣoḍaśabhyaḥ..., aṣṭādaśabhyaḥ..., viṁśatyai svāhā...

Then suddenly the author leaps to 98 and 100...*aṣṭānavatyai svāhā śatāya svāhā.* The numbers after 100 are left to the imagination of the students, since, given the method of adding 2 to the preceding number, the rest of the numbers after 100 could be easily found out.

This represents a series of arithmetic progression and of the multiples of 2.

TS.7.2.14 is a repetition of TS.7.2.12 above, with the omission of the first term, *ekasmai svāhā; it starts with tribhyaḥ svāhā* and ends with *ekānnaśatāya svāha* so far as the series of arithmetic progression is concerned. The last *śatāya svāhā* introduces the first word of the next series and rank.

TS.7.2.15 states the technique of finding out (i) the multiples of 4 by the addition of 4 to each preceding number and (ii) by the addition of two previous odd numbers (like 1+3 = 4; 3+5 = 8; 5+7 = 12 etc.) given in TS. 7.2.12. It thus repeats the technique stated by VS. 18.25 discussed before. The passage is as follows:

caturbhyaḥ svāhā, aṣṭābhyaḥ..., dvādaśabhyaḥ..., ṣoḍaśabhyaḥ..., viṁśatyai..., and then a sudden jump to 96 as *ṣaṇṇavatyai svāhā, śatāya svāhā.*

This is a series of arithmetic progression with a difference of +4.

TS.7.2.16 gives (i) the multiples of 5 and (ii) the series of arithmetic progression with a difference of +5.

pañcabhyaḥ svāhā, daśabhyaḥ..., pañcadaśa–bhyaḥ..., viṁśatyai svāhā and then suddenly jumping to 95, *pañcanavatyai svāha, śatāya svāhā.*

TS.7.2.17 gives (i) the multiples of 10 and (ii) the series of arithmetic progression with a difference of +10. Thus we have,

daśabhyaḥ svāhā, viṁśatyai..., triṁśate..., catvāriṁśate..., pañcāśate..., ṣaṣṭyai..., saptatyai..., aśītyai..., navatyai..., śatāya svāhā.

TS.7.2.18 mentions (i) the multiples of 20 and (ii) the arithmetic progression with a difference of 20. It runs as follows:

viṁśatyai svāhā, catvāriṁśate..., ṣaṣṭyai..., aśītyai..., śatāya svāhā.

TS.7.2.19 gives a peculiar series which cannot be called either arithmetic progression or geometric progression, but is a strange combination of both. It is as follows:

pañcāśate svāhā..., śatāya..., dvābhyam śatābhyām..., tribhyaḥ śatebhyaḥ... caturbhyaḥ śatebhyaḥ..., pañcabhyaḥ śatebhyaḥ..., ṣaḍbhyaḥ śatebhyaḥ..., saptabhyaḥ śatebhyaḥ..., aṣṭābhyaḥ śatebhyaḥ..., navabhyaḥ śatebhyaḥ..., sahasrāya svāhā.

Translated into number-symbols, the series is: 50, 100, 200, 300, 400, 500, 600, 700, 800, 900, 1000.

Excluding the first number 50, the series is a regular arithmetic progression series with a difference of 100. Thus, 100, 200......1000.

If we include the initial number 50, the series can be re-written in different form such as–

50 = 50 x 1

100 = 50 x 2

200 = 50 x (2x2)

300 = 50 x (2x3)

400 = 50 x (2x4)

500 = 50 x (2x5)

600 = 50 x (2x6)

700 = 50 x (2x7)

800 = 50 x (2x8)

900 = 50 x (2x9)

and 1000 = 50 x (2x10)

This gives us that the series is enumerating the product of 50 with numbers 2 and its multiples. In spite of different transformations of the series, the series does not give us a consistent formula in which it can be fitted. If we omit the number

50, we obtain a regular formula. We do not know the propriety or purpose of the inclusion of the number 50 in the beginning.

We can also interpret the series from an altogether different point of view, which may be called a geometrical interpretation. We write the number 50 as a sum of two squares. There are again two possibilities. $50 = 7^2+1^2$ or $50 = 5^2+5^2$. If, suppose, we take the numbers 7 and 1, or 5 and 5 as the length of the two sides viz. a and b of a right-angled triangle with the hypotenuse c, the number 50 gives us the square of the hypotenuse c;

$\therefore 50 = c^2$;

and we have the equation,

$a^2+b^2 = c^2$ (by the Śulba-sūtra or Pythagoras theorem)

When translated into number, the equation is:

(7^2+1^2) or $(5^2+5^2) = 50$

The length of the hypotenuse, therefore, will be equal to $\sqrt{50}$. This gives the value of $\sqrt{50}$ which otherwise is difficult to obtain by ordinary procedure of finding out square roots.

Coming to the series proper, we re-write the series as:

$50 = (7^2+1^2)$ or (5^2+5^2)

$100 = 2\ (7^2+1^2)$ or $2\ (5^2+5^2)$

$200 = 4\ (7^2+1^2)$ or $4\ (5^2+5^2)$

$300 = 6\ (7^2+1^2)$ or $6\ (5^2+5^2)$

$400 = 8\ (7^2+1^2)$ or $8\ (5^2+5^2)$

$500 = 10\ (7^2+1^2)$ or $10\ (5^2+5^2)$

$600 = 12\ (7^2+1^2)$ or $12\ (5^2+5^2)$

$700 = 14\ (7^2+1^2)$ or $14\ (5^2+5^2)$

$800 = 16\ (7^2+1^2)$ or $16\ (5^2+5^2)$

$900 = 18\ (7^2+1^2)$ or $18\ (5^2+5^2)$

and $1000 = 20\ (7^2+1^2)$ or $20\ (5^2+5^2)$ and so on.

If we put *a* for any of the right hand side expressions viz. either (7^2+1^2) or (5^2+5^2), the picture of the series that we get is as follows:

$50 = 1a$

$100 = 2a$

$200 = 4a$

$300 = 6a$

$400 = 8a$

$500 = 10a$

$600 = 12a$

$700 = 14a$

$800 = 16a$

$900 = 18a$

and $1000 = 20a$

This series, excluding the first expression (viz. $50 = 1a$), turns out to be a series of arithmetic progression with a difference of 2.

The purpose of the series seems to be to guide to get the square-root values, in terms of a concrete length, of the awkward numbers like 200, 300, 500, 600, 700, 800 and 1000 whose square root cannot be easily found out.

It is to be noted that the equation $50 = 5^2+5^2$ gives us an isosceles right-angled triangle.

Still the purpose behind stating this odd series is not clear; it may have been stated for geometrical purposes.

The theorem $a^2+b^2 = c^2$, where *a* and *b* are the sides of a right-angled triangle and *c* is the hypotenuse is known as of Pythagoras.

A.K. Bag (*ibid.* p. 123, 165), however states that it was already enunciated by the *Śulbasūtrakāra* named Baudhāyana (c. 600 B.C.).

18.2. Geometric Progression

Geometric progression is defined as "A series of numbers in which each, after the first, is the product of the preceding number and a fixed number, called the *common ratio* .

Examples: 1, 2, 4, 8, 16, 32, 64.....(common ratio = 2); 1, 3, 9, 27, 81, 243...(common ratio = 3).

The nth term of a geometric progresion is given by the formula

$l = ar^{n-1}$

where l = nth term, a = first term, r = common ratio, n = number of terms in the series (upto and including the nth term)".[34]

There is only one, solitary example in the whole of the Vedic literature of the series of geometric progression and it is the verse VS.17.2 which is repeated in part in TS.4.4.11.4 and KS. 17.10. The verse is quoted before. The verse in question quotes the following numbers:

eka = 1	$= 10^0$
daśa = 10	$= 10^1$
śata = 100	$= 10^2$
sahasra = 1000	$= 10^3$
ayuta = 10,000	$= 10^4$
niyuta = 100,000	$= 10^5$
prayuta = 1000,000	$= 10^6$
arbuda = 10,000,000	$= 10^7$
ny-arbuda = 100,000,000	$= 10^8$

samudra = 1000,000,000 = 10^9

madhya = 10,000,000,000 = 10^{10}

anta = 100,000,000,000 = 10^{11}

parārdha = 1000,000,000,000 = 10^{12}

We will find that any pair of two consecutive members in this series constantly maintains the ratio of 10. Thus, in the case of the first two members, 1 and 10, we have the ratio of 10:1; in the case of the next pair, viz. 10 and 100, we have the same ratio of 10:1 and so on. This series also has the potentiality of being infinitely expanded to infinity.

Even if differently interpreted as it can be in the following way, the series has still the status of one in geometric progression. It is in the following way:

ekā ca...daśa ca = 1+10 = 11

daśa ca śatam ca = 10+100 = 110

śatam ca sahasram ca = 100+1000 = 1100

sahasram ca ayutam ca = 1000+10,000 = 11,000

ayutam ca niyutam ca = 10,000+100,000 = 1,10,000

niyutam ca prayutam ca = 100,000+1000,000 = 1,100,000

prayutam ca arbudam ca = 1000,000+10,000,000 = 11,1000,000

arbudam ca ny-arbudam ca = 10,000,000+100,000,000 = 110,000,000

ny-arbudam ca samudraś ca = 100,000,000+1,000,000,000 = 1,100,000,000.

samudraś ca madhyam ca = 1,000,000,000+10,000,000,000 = 1,1000,000,000

madhyaṁ ca antaś ca = 10,000,000,000+100,000,000,000 = 110,000,000,000.

antaś ca parārdhaś ca = 100,000,000,000+1,000,000,000,000 = 1,100,000,000,000.

The particle *ca* signifies addition. In each of the pair of two consecutive numbers in the above series there is the constant ratio of 10:1, and hence the series is in geometric progression. It is also to be noted that this series also provides an example of infinite series.

All the above series satisfy the general formula of geometric progrossion, viz.

$x, x^2, x^3, x^4 \ldots x^n \ldots$ *ad infinutum.*

The same series of geometric progression is given in TS.7.2.20.

An interesting point may be noted in this connection. The verses from VS.18.24 and 25 end with, what we may call as a kind of refrain phrase viz. *yajñena kalpantām*. This phrase is used at the end of all the verses from the *adhyāya* VS.18, which is popularly known among the circles of Vedic reciters as *camakādhyāya*, since it contains the words *ca* and *me* at every step.

The passages from TS.7.2.11-20 contain at the end the phrase *sarvasmai svāhā.*

Apart from the literal meaning of the two phrases, viz. 'Let it result by *yajña* (for *yajñena kalpantām*)' and '*svāhā* to all (for *sarvasmai svāhā*),' the real meaning hidden in the phrases seems to be technical, viz. 'and in this way proceed *ad infinitum*' or 'and so on' or 'etcetera', as we may use in modern works. To use such apparently meaningless refrain phrases to show the way to further mathematical facts or results seems to be one of the many devices in Vedic times. Such phrases also imply that the previous examples or series or enumeration are based on certain mathematical principles or methods, and also that the students are, or should be, fully acquainted with them and should proceed on the lines indicated or implied in them. The practice of the Vedic theorists is to illustrate the mathematical method by giving examples upto the nth term and then say '*yajñena kalpantām*' or '*sarvasmai svāhā*' or some such like. A collection and study of all such technical terms and devices is worth-pursuing. Such terms discard the word-meanings and adopt a purely technical meaning. Unfortunately, we do not find such terms clearly defined in any of the Vedic texts or in the commentaries thereon. The possible reason seems to be that such technical terms might have become so well-known and current in the times that it was found not necessary to define them; their meaning must have been very clear to them. This line of reasoning pre-supposes a long pre-Vedic tradition in the field,–so old that even Vedas might look modern in comparison with that. There must have been a long *guru-śiṣya paramparā* (teacher-pupil-tradition) in which such interpretational principles were transmitted through oral tradition and hence were not included in the texts. The tradition being oral, the task of explaining the different interpretational and mathematical principles and devices of mathematical operations was left to the teacher or *guru*.

19

Zero

The concept of zero in mathematics by the Vedic mathematicians is one of the marvellous invention which has very few parallels in the history of other sciences. The introduction of zero in the place-notation has changed the whole course of mathematics in the later, post-Vedic times. It has enabled man to build up a number-system which can be extended to any limit, even upto infinity. This number-system based on the device of zero enabled him to measure and count infinitely. With the invention of zero, mathematics entered into a new phase and became the most important tool in the hands of man by which he succeeded to measure, compute, count and cover up the whole realm of knowledge where compupation is necessary.

In mathematics the concept of zero is used mainly in two ways:[35] (i) as a number as in 0, 1, 2, 3 ...etc. and (ii) as a value of a variable. As a number, with which we are mainly concerned here it is again of two types: (i) it is "an immediate predecessor" of the positive number 1. This zero is the lowest limit of the positive numbers and the highest limit of the negative number -1, -2, -3, etc. (ii) as a marker of an empty place in numbers like 10, 20, 30 etc.

This zero, as a number, is a concept and is totally different from the symbol 0 which is used to indicate an empty, un-occupied place in numbers like 10, 20, 105, 3047 etc. The zero as a marker of an empty space is only a substitute for a vacuum in place-notation. One zero marks one vacant place, two zeroes mark two vacant places, three zeroes marks three vacant places and so on. Thus in 10, we have one vacant place; in 100 we have two and so on. Basically, therefore, the two types of zeroes function differently. The number zero in the series 0, 1, 2, 3 etc. does not requires a place-notation; the symbol zero in 10, 20 etc. requires the place-notation. It cannot be comprehended without the place-notation. The former is a real number; the latter is a symbol. Actually the two should have been written with different symbols. The former zero as a number preceding one is obtained by the process of subtraction, as 1-1, or more generally, x-x. The latter zero as a symbol of an empty space cannot be obtained by the process of subtraction. Or is it also obtained by subtraction? But then what is that number from which it is subtracted? We will discuss all these problems later on.

Be it as it is, we can classify the zero in the following way:

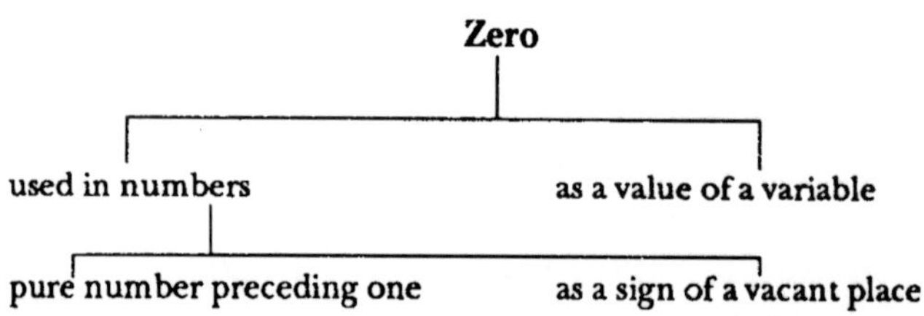

The system of the notation of numbers in terms of their place-value in the decimal number-system has proved so convenient, consistent and useful that it has thrown all other number-systems not only into back-ground but in total oblivion.

19.1. Zero and Śūnya

Since the two types of mathematical zero are functionally different, we should prefer to call them by different terms. We call the first type of zero, which, as a number, precedes +1 and succeeds—1 by the term 'zero' itself. We prefer to call the other type of zero which, as an indicator of an un-occupied place in, say, 10, means 'void, nothingness' etc. by the term *Śūnya*. It must be remembered, however, that though 'zero' and '*Śūnya*' are two independent words from two different languages viz. English and Sanskrit, they both linguistically mean the same thing, viz. 'void, vacant, nothingness, absence' etc. Not only this, but, as we shall see later on, the English word 'zero' is philologically derived from the Sanskrit word '*Śūnya*' Robert W. Marks[36] rightly maintains this distinction and defines the two words separately. Yet, since both are symbols, it is not necessary to follow the distinction between *Śūnya* and zero so strictly. For our present study, therefore, *Śūnya* = zero.

19.2. Zero in other sciences

Before we attempt to explain why and how the digital place in the number 10 falls vacant and how it is occupied–or rather seems to be occupied–by the symbol zero, it is useful we try to search out similar phenomenon, if any, in other sciences–other than mathematics. If we find such a phenomenon in any other science, we have a lot of ground and scope to compare the two and come to certain conclusions which may throw light on the principles and techniques of the Vedic science of mathematics.

We have seen above that the zero in 10 can be termed as *Śūnya*, as against the other 'zero'. This immediately suggests the likely Indian sciences which may provide us the possibility of comparison so far as the concept of 'void, nothingness, absence' etc. is concerned. And these sciences are only three, viz. the buddhistic philosophy or *bauddha darśana* which propounds the theory of *Śūnyavāda*; the science of Indian Logic or *Nyāya darśana* or rather, *Nyāya-Vaiśeṣika darśana* and the science of Indian grammar or *Vyākaraṇa Śāstra* whose greatest master is Pāṇini. The 'vacuum or void' in Buddhistic philosophby is called *Śūnya*. The 'vacuum or void' in Nyāya-philosophy is called '*abhāva*' (=*a*+*bhāva*

= not+*existstence* = non-existence or *absence*) i.e. 'negation'. And the vacant place in Pāṇini's grammar is signified by the term '*lopa*' (lit-elision, non-appearance).

We will examine one by one the different view-points of these three sciences which have considered the vacuum or zero in their own way.

19.2.1. *The Buddhistic Śūnya*

Let us start with the philosophical concept of *Śūnyavāda*, 'doctrine of nihilism' as S. Radhakṛshnan calls it,[37] expounded by Buddhism. Whether Gautam Buddha wrote anything about *Śūnyavāda* philosophy which is said to be authored by him, or not–we do not know. But, as the tradition goes, he did not write anything. What he did was to preach orally whatever doctrines he cherished; and that too, not in Sanskrit but, in Pāli language which was the spoken language of the masses in those times. We, therefore, have no choice but to search the Pāli literature to find out the tenets of the *Śūnyavāda*-philosophy.

The only Pali text which is available for the exposition of *Śūnyavada* is *Milindapañhapāli.* [38]

Another text which is in Sanskrit and which expounds and elaborates the doctrine of *Śūnyavāda* is by Nāgarjuna;[39] its title is *MadhyamakaŚāstra* which tries to give a strong logical and philosophical foundation to the whole theory.

The exposition and the discussion of *Śūnyavāda* which is given here is based upon these two books.

19.2.1.1. *Milindapañhapāli*

As the title indicates, (which literally means 'the series of questions of King Milinda'), the book is written in the form of questions and answers between King Milinda and Nāgasena. Milinda is generally identified with the Greek king Menander who was the ruler of Ghazni and adjoining areas of Kabul valley.[40]

Historically both King Milinda and Nagasena are placed in the first century[41] B.C. King Milinda puts the questions and Nagasena answers them. The text which is traditionally supposed to be the basis of later Śūnyavāda-philosophy forms the beginning part of Ch.2 of *Milindapañha*, called *Mahāvaggo* with the subsection 1 entitled as *Rathūpamāya Puggalavīmaṁsanam.* The text is quoted below for ready reference.[42]

१. अथ खो मिलिन्दो राजा येनायस्मा नागसेनो तेनुपसङ्कमि। उपसङ्कमित्वा आयस्मता नागसेनेन सद्धिं सम्मोदि। सम्मोदनीयं कथं सारणीयं वीतिसारेत्वा एकमन्तं निसीदि। आयस्मापि खो नागसेनो पटिसम्मोदनीयेनेव मिलिन्दस्स रञ्ञो चित्तं आराधेसि।

अथ खो मिलिन्दो राजा आयस्मन्तं नागसेनं एतदवोच—"कथं भदन्तो ञायति—किंनामोसि, भन्ते" ति? "नागसेनो ति खो अहं, महाराज, ञायामि। 'नागसेनो ति खो मं, महाराज, सब्रह्मचारी समुदाचरन्ति। अपि च मातापितरो नामं करोन्ति—'नागसेनो' ति वा 'सूरसेनो' ति वा 'वीरसेनो' ति वा 'सीहसेनो' ति वा। अपि च खो, महाराज, सङ्खा समञ्ञा पञ्ञत्ति वोहारो नाममत्तं यदिदं नागसेनो' ति। न हेत्थ पुग्गलो उपलब्भती" ति।

अथ खो मिलिन्दो राजा एवमाह—"सुणन्तु मे, भोन्तो पञ्चसता योनका, असीतिसहस्सा च भिक्खू! अयं नागसेनो एवमाह—'न हेत्थ पुग्गलो उपलब्भती' ति। कल्लं नु खो तदभिनन्दितुं" ति? अथ खो मिलिन्दो राजा आयस्मन्तं नागसेनं एतदवोच—"सचे, भन्ते नागसेन, पुग्गलो नूपलब्भति, को चरहि तुम्हाकं चीवरपिण्डपातसेना-सनगिलानप्पच्चयभेसज्ज- परिक्खारं देति? को तं परिभुञ्जति? को सीलं रक्खति? को भावनमनुयुञ्जति? को मग्गफलनिब्बानानि सच्छिकरोति? को पाणं हनति? को अदिन्नं आदियति? को कामेसु मिच्छाचारं चरति? को मुसा भणति? को मज्जं पिवति? को पञ्चानन्तरियकम्मं करोति? तस्मा नत्थि कुसलं, नत्थि अकुसलं, नत्थि कुसलाकुसलानं कम्मानं कत्ता वा कारेता वा, नत्थि सुकतदुक्कटानं कम्मानं फलं विपाको। सचे, भन्ते नागसेन, यो तुम्हे मारेति, नत्थि तस्सापि पाणातिपातो। तुम्हाकं पि, भन्ते नागसेन, नत्थि आचरियो, नत्थि उपज्झायो, नत्थि उपसम्पदा। 'नागसेनो ति मं, महाराज, सब्रह्मचारी समुदाचरन्ती' ति यं वदेसि, कतमो एत्थ नागसेनो"?

"किन्नु खो, भन्ते केसा नागसेनो" ति? "न हि, महाराजा" ति। "लोमा नागसेनो" ति? "न हि, महाराजा" ति।

"नखा... दन्ता... तचो... मंसं... न्हारू... अट्ठि... अट्ठिमिज्जं... वक्कं...हदयं... यकनं... किलोमकं... पिहकं...पप्फासं...अन्तं... अन्तगुणं... उदरियं... करीसं...पित्तं.... सेम्हं... पुब्बो... लोहितं.. .सेदो... मेदो... अस्सु... वसा... खेळो... सिङ्घाणिका... लसिका... मुत्तं... मत्थके मत्थलुङ्गं नागसेनो" ति? "न हि, महाराजा" ति। "किन्नु खो, भन्ते, रूपं नागसेनो" ति? "न हि, महाराजा" ति। "वेदना नागसेनो" ति? "न हि, महाराजा" ति। "सञ्ञा नागसेनो" ति? "न हि, महाराजा" ति। "सङ्खारा नागसेनो" ति? "न हि, महाराजा" ति। "विञ्ञाणं नागसेनो" ति? "न हि, महाराजा" ति। "किं पन, भन्ते, रूपवेदनासञ्ञासङ्खारविञ्ञाणं नागसेनो" ति? "न हि, महाराजा" ति। "किं पन, भन्ते, अञ्ञत्र रूपवेदनासञ्ञासङ्खारविञ्ञाणं नागसेनो" ति? "न हि, महाराजा" ति। "तमहं, भन्ते, पुच्छन्तो पुच्छन्तो न पस्सामि नागसेनं। नागसेनसद्दो येव नु खो, भन्ते, नागसेनो" ति? "न हि, महाराजा" ति। "को पनेत्थ नागसेनो? अलिकं त्वं, भन्ते, भाससि मुसावादं, नत्थि नागसेनो" ति?

अथ खो आयस्मा नागसेनो मिलिन्दं राजानं एतदवोच—"त्वं खोसि, महाराज, खत्तियसुखुमालो अच्चन्तसुखुमालो। तस्स ते, महाराज, मज्झन्हिकसमयं तत्ताय भूमिया उण्हाय वालिकाय खराय सक्खरकथलिकाय मद्दित्वा पादेनागच्छन्तस्स पादा रुज्जन्ति, कायो किलमति, चित्तं उपहञ्ञति, दुक्खसहगतं कायविञ्ञाणं उप्पज्जति। किन्नु खो त्वं पादेनागतोसि, उदाहु वाहनेना" ति? "नाहं भन्ते, पादेनागच्छामि। रथेनाहं आगतोस्मी" ति।

"सचे त्वं, महाराज, रथेनागतोसि, रथं मे आरोचेहि। किन्नु खो, महाराज, ईसा रथो" ति? "न हि, भन्ते" ति। "अक्खो रथो" ति? "न हि, भन्ते" ति। "चक्कानि रथो" ति? "न हि, भन्ते" ति। "रथपञ्जरं रथो" ति? "न हि, भन्ते" ति। रथदण्डको रथो" ति? "न हि, भन्ते" ति। "युगं रथो" ति? "न हि, भन्ते" ति। "रस्मियो रथो" ति? "न हि, भन्ते" ति। "पतोदलट्ठि रथो" ति? "न हि, भन्ते" ति। "किन्नु खो, महाराज, ईसा-अक्ख-चक्क-रथपञ्जर-रथदण्ड-युग-रस्मि-पतोदा रथो" ति? "न हि, भन्ते" ति। "किं पन, महाराज, अञ्ञत्र ईसाअक्खचक्करथपञ्जररथदण्डयुगरस्मिपतोदा रथो" ति? "न हि, भन्ते" ति। "तमहं, महाराज पुच्छन्तो पुच्छन्तो न पस्सामि रथं। रथसद्दो येव नु खो, महाराज, रथो" ति? "न हि भन्ते" ति। "को पनेत्थ रथो? अलिकं त्वं, महाराज, भाससि मुसावादं, नत्थि रथो। त्वंसि, महाराज सकलजम्बुदीपे अग्गराजा। कस्स पन त्वं

भायित्वा मुसावादं भाससि? सुणन्तु मे, भोन्तो पञ्चसता योनका, असीतिसहस्सा च भिक्खू! अयं मिलिन्दो राजा एवमाह—'रथेनाहं आगतोस्मी' ति। 'सचे त्वं, महाराज, रथेनागतोसि, रथं मे आरोचेही' ति वुत्तो समानो रथं न सम्पादेति। कल्लं नु खो तदभिनन्दितुं" ति? एवं वुत्ते पञ्चसता योनका आयस्मतो नागसेनस्स साधुकारं दत्वा मिलिन्दं राजानं एतदवोचुं—"इदानि खो त्वं, महाराज, सक्कोन्तो भासस्सू" ति।

अथ खो मिलिन्दो राजा आयस्मन्तं नागसेनं एतदवोच—"नाहं, भन्ते, नागसेन, मुसा भणामि। ईसं च पटिच्च, चक्कानि च पटिच्च, रथपञ्जरं च पटिच्च, रथदण्डकं च पटिच्च 'रथो' ति सङ्खा समञ्ञा पञ्ञत्ति वोहारो नाममत्तं पवत्तती" ति।

"साधु खो त्वं, महाराज, रथं जानासि। एवमेव खो, महाराज मय्हं पि केसे च पटिच्च, लोमे च पटिच्च...मत्थके मत्थलुङ्गं च पटिच्च, वेदनं च पटिच्च, सञ्ञं च पटिच्च, सङ्खारे च पटिच्च, विञ्ञाणं च पटिच्च, 'नागसेनो' ति सङ्खा समञ्ञा पञ्ञत्ति वोहारो नाममत्तं पवत्तति। परमत्थतो पनेत्थ पुग्गलो नूपलब्भति। भासितं पेतं, महाराज, वजिराय भिक्खुनिया भगवतो सम्मुखा—

यथा हि अङ्गसम्भारा, होति सद्दो रथो इति।

एवं खन्धेसु, सन्तेसु, होति सत्तो ति सम्मुती" ति।।

"अच्छरियं, भन्ते नागसेन, अब्भुतं भन्ते नागसेन! अतिचित्रानि पञ्हपटिभानानि विसज्जितानि। यदि बुद्धो तिट्ठेय्य साधुकारं ददेय्य—'साधु साधु, नागसेन, अतिचित्रानि पञ्हपटिभानानि विसज्जितानी" ति।

The Buddhistic philosophy regarding *Śūnyavāda* as expounded in the above passage may be summed up in nutshell as follows:

If we see a thing, we perceive that the thing is made of parts. Take, for example, the chariot, which is Nagasena's example. The chariot is made up of the seat, the covering canopy, the wheels, the yoke etc. None of these parts can be pinned down as 'the chariot'. If we take away all the parts, no such thing as a chariot remains. In other words, the chariot is born or made up from nothing; the wheels, the yoke etc. are all its parts and not the chariot. The chariot, to be a chariot, requires all these parts; these parts in turn require some agent to assemble them together into the form of a chariot. The agent again requires some other instruments like the wood which again comes from tree, the sharpening instruments, the axe etc. This all means that neither the chariot nor its making is independently done; it requires the means and the help from outside. It by itself cannot be its own cause of creation. The universe also in the same way cannot be created by any individual or single entity. And if we exclude all these causes which are said to be instremental in creating the universe, what we get at the end or basis is 'nothingness' and not a positive entity. The philosophy thus comes to the conclusion that there is nothing at the root of this universe which can be pointed out to be the single, basic principle or cause. There is total vaccum or *Śūnya*.

Represented in equational form, the whole process of argumentation and explanation will look like the following:

chariot = the seat+the covering canopy+the yoke+the wheels etc.

putting symbols like x,y,z etc. we have,

chariot = $x+y+z+a$...etc.

If now we subtract one by one each part of the chariot, what we get is–

chariot = ($x+y+z+a$...etc.) - ($x+y+z+a$..etc.) = 0

That is to say, that the chariot is reduced to total zero if we subtract all its parts which make it.

This type of *Śūnya* in Buddhism is only comparable to the zero in mathematics which is obtained by subtracting 1 from 1 or more generally x-x and not the zero in the number-symbol 10 which symbolizes an empty, un-occupied space. It is also to be noted that this *Śūnya* is available only by the process of subtraction.

Actually the passage quoted above is traditionally regarded as expounding what they call as *anattāvāda* (SK. *anātmavāda*) and not *śūnyavāda*. *anattāvāda* means that there is no *attā* i.e. *ātman*, 'soul' as such; everything is destructible and there is no such thing as indestructible *attā*. Yet, the passage given here serves as the basis for the later *śūnyavāda* developed by the *mādhyamika* school. The famous upaniṣadic phrase '*neti neti*' (= not this, not this) finds its echo in the *anattāvāda* of *Milindapañhapāli*.

Nāgārjuna is the next Buddhist philosopher to expound the philosophy of śūnyavāda in better philosphical terms. The whole argument of Nāgārjuna can be briefly stated as follows: any creation requires the three causes, viz. *kartā*, *karaṇa* and *samavāyikāraṇa*. Thus in the creation of the *ghaṭa*, earthen pot, the three factors viz. the pot-maker, the earth (*mṛttikā*) and the stick and the wheel are required. In other words, even though the pot-maker is said to be the *kartā* of the pot, he by himself is incapable of manufacturing the pot without the help of the relevant material viz. the earth, the stick, the wheel, water etc. His *kartṛtva* is thus *sāpekṣa* i.e. relative to the other auxiliary material. Since what is *sāpekṣa* i.e. relative in theory and practice is not the truth and has no independent existence by itself (as in the case of the illustration of pot) so also in the case of the whole world, its original cause, by whatever name one may call it, *brahman*, *Iśvara*, *kāla*, *prakṛti* or *paramāṇu*, is not independent in the sense that it has any power to create the world by itself and requires no help from outside of itself. This is what is called by *pratītya samutpāda* or *bhāvānām nissvabhāvatva* by Nāgārjuna. If this logic is correct, it leads us to the next conclusion viz. that there is no independent cause at the root of the whole creation; in other words, there cannot be any single, independent entity which can be cited as the root-cause of this vast creation. *cf. Mādhyamakaśātra*, 17.31: *yathā nirmitakam śāstā nirmimītarddhisampadā/nirmito nirmimītānyam sa ca nirmitakaḥ punaḥ//* also 17.32: *tathā nirmitakākāraḥ kartā karma ca tatkṛtam/tad yathā nirmitenānyo nirmito nimitas tathā//* Nāgārjuna, therefore, says (*cf.*1.10):

anālambana evāyam san dharma upadiśyate/athānālambane dharme kuta ālambanam punaḥ.

Then Nāgārjuna comes to the conclusion of *śūnyatā 'zero-ness'* as *pratītya-samutpāda; cf.* *Madhyama-śastra.* [42A] *yaḥ pratītyasamutpādaḥ śūnyatām tām pracakṣmahe.* The commentator Candrakīrti explains the term *pratītya-samutpāda* as: *yo'yaṃ pratītya-samutpādaḥ hetupratyayānapekṣyaḥ aṅkuravijnānādi- prādurbhāvaḥ, sa svabhāvena anutpādaḥ. yaśca svabhāvena anutpādo bhāvānām sā śūnyatā.* He quotes from different texts; *cf.* a quotation from *Āryalaṅkāvatāra*: *svabhāvānutpattim saṁdhāya mahāmate sarvadharmāḥ śūnyāḥ iti mayā deśitāḥ; cf.* also one from the text entitled *Dvyardhaśatikā*: *śūnyāḥ sarvadharmāḥ nissvabhāvayogena.* No single independent entity possesses independently the power to create the world; everything in the world is devoid of such independent power. This is the *nissvabhāvatva* or *naissvābhāvya* of the entity. Such a state of being devoid of *svabhāva* is in fact what is meant by the term *śūnyavāda* in the system of the *Mādhyamika* school.

The whole argument clearly means that Nāgārjuna does not accept *śūnya* in the sense of total 'void or nothingness' as the root-cause of the world, but his *śūnya* refers to the fact that there is no single, independent principle at the root of the creation, which by its inherent nature is omni-potent, omni-scient and omni-present. This is made very clear by the commentator Candrakīrti on the verses 14.11, 12, 13: *abhāvārtham hi śūnyatārtham āropya prasaṅga udbhāvito bhavatā. na ca vayam abhāvārtham śūnyatārtham vyācakṣmahe, kim tarhi pratītyasamutpādārtham.* This clearly refutes the idea of the *śūnya* as 'the total void, nothingness or absence' - What the *śūnyavāda* refers to seems to be *svabhāva-śūnyatā,* 'incapacity by its inherent nature' of a single, independent principle to be the root-cause of the creation.

The *śunyavāda* of the *Mādhyamika* school of Buddhism, however, is variously interpreted and understood or misunderstood.[42B] And, on the authority of Nāgārjuna and his commentators themselves, they interpret *śūnya* to mean 'a positive entity'.

The *śūnya* of the *śūnyavādī mādhyamika* school of Buddhism thus swings between total void and the positive entity.

If, therefore, the *śūnyavāda* does not mean and deal with *śūnya* in the sense of void, nothingness or absence (*abhāva*), but means or refers to positive entity, it is of no use to us from mathematical point of view so far as its comparison with the mathematical zero is concerned. Even if we take Nāgārjuna's *śunya* to mean void, in the philosophical sense or zero in the mathematial sense, it does not serve any purpose for us as mathematicians because the *śūnyavāda* does not state anywhere any formal procedure, as in the mathematical technique, to arrive at the zero, or does not enunciate any positional analysis or notation, as in mathematics, which can account for the presence of an unoccupied or vacant place.

19.2.2. *The abhāvavāda of the Naiyāyikas*

There is yet another system of Indian philosophy, viz. *Nyāya-śāstra,* which treats the concept of zero. The zero in this system is called *abhāva*. *abhāva,* according to Śivāditya (c. 984 AD) is of four kinds: *abhāvas tu prāgabhāva-pradhvaṁsābhāva-atyantābhāva-anyonyābhāvalakṣaṇaḥ caturvidhaḥ.* [43] Annambhaṭṭa in his *Tarkasaṁgraha* (*ibid.* p.62) defines the four types of *abhāvas* as follows:-

19.2.2.1. *prāg-abhāva*: *anādiḥ sāntaḥ prāgabhāvah . utpatteḥ pūrvam kāryasya.* The *prāg-abhāva* is *anādi*, 'without beginning', yet *sānta*'has as end' and exists before the *kārya* (i.e. production of an effect). This is called 'antecedent negation'.

19.2.2.2. *pradhvaṁsābhāva: sādir anantaḥ pradhvaṁsaḥ* utpattyanantaram kāryasya. The *pradhvaṁsābhāva* has a beginning but no end; and it refers to the time after the *kārya* i.e. production of the effect. It is called 'consequent negation.'

19.2.2.3. *atyantābhāva*: *traikclikasaṁsargāvachinnapratiyogitākaḥ atyantābhāvaḥ yathā bhūtale ghaṭo nāsti iti.* This is called an 'absolute negation'.

19.2.2.4. *anyonyābhāva: tādātmyasaṁbandhā vacchinnapratiyogitākaḥ anyonyābhāvaḥ. yathā ghaṭaḥ paṭo na bhavati iti.* This is called a 'reciprocal negation.'

The first two types form one group of what may be called 'transient negation'; and the other two form a group of what may be called 'permanent negation'.

The Nyāyamañjarī (*Prathamāhnika)* defines the *abhāva* as follows : *na hi niḥśeṣasāmarthyarahittattvam abhāvalakṣaṇam, api tu nāstīti jñānagamyatvam. satpratyayagamyo hi bhāva iṣyate, asatpratyaya-gamyas tu abhāva iti.* The *Siddhānta Candrodaya,* a commentary on *Tarkasaṁgraha* by Srikṛṣṇa Dhūrjaṭi says : *pratiyogijñānādhīnatvam abhāvattvam.* The *Siddhānta Muktāvali* of *Viśvanātha Pañcānana* explains *abhāva* as: *drvyādiṣaṭkānyonyābhāvatvam.* The *Sarvadarśana-saṁgraha* in its exposition of *Nyāya* - theory states the following definition of *abhāva* : *asamavāyatve saty asamavāyitvam.* All these definitions, however, are not absolute definitions of *abhāva* and are framed in relation to *bhāva;* or existence. All these definitions of *abhāva* in *Nyāya* also donot serve our purpose of comparison, first, because they are all relative to *bhāva,* secondly like the *mādhyamika śūnyavāda* , donot lay down any formal procedure to arrive at *abhāva.* Thirdly, in both, the *mādhyamika* and *Nyāya* schools, the *śūnyavāda* and the *abhāvavāda,* are principles and not technique unlike in mathematics. The *śūnyavāda* refers to *śūnya* as the highest principle at the root of the universe; the *abhāvavāda* takes *abhāva* as one of the seven basic entities and not a technique; *cf. Tarkasaṁgraha; dravyaguṇakarmasāmanya-viśeṣasamavāyābhāvāḥ sapta padārthāḥ.* The *abhāva* in the *Tarkasaṁgraha,* therefore, can at the most compare only with one of the *Pāṇinian* six *vyavasiṭas* and not *lopa* or zero.

The chief characteristics of the *abhāva* - zero of the Naiyāyikas are mainly two: (i) the concept of *abhāva* is related to *bhāva* i.e. existence; and (ii) its classification into four type seems to have been based upon the existence or non-existence of the two conditions of *ādi* and *anta*. The four types are: *anādi-sānta* (without beginning but with an end) which is *prāgabhāva; sādi-ananta* (with beginning but without end) which is *pradhvaṁsā-bhāva*; *anādi-ananta* (without beginning and without end) which is *atyantābhāva*; this is absolute negation; at this stage there is a sudden shift of the basis of classification Instead of having the fourth type as *sādi* and *sānta*, the fourth type is totally different depending or based on the reciprocal relation of two entities.

Of these four types, the third one viz. *antyantābhāva*, which is absolute and does not refer to any non - *abhāva* entity, is beyond comprehension, since in that state of absolute non-existence, nothing remains; there lives nobody even to comprehend the knowledge of *abhā*va.

abhāva is reckoned for the first time as an independent category by the Vaiśeṣikas after Praśastapāda. Then we find it in Udayana's *Kiraṇāvali*. Śivāditya's *Saptapadārthī* includes it as the seventh *padārtha*. As D. Gurumurti puts it in his introduction to Śivāditya's *Saptapadārthī*.

"Abhāva arose as a logical concept. It is serviceable for intellectual distinction. In knowledge, the idea of negation as the counterpart of affirmation is necessarily involved. All idealistic systems of philosophy are based on the opposition between the knowable and the unknowable. As relations play a large part in the intellectual explanation of the universe, they are distinguished from that which is above all relational consciousness. "When we speak of a thing the fact of its being or existence is emphasised; while when we speak of a relation its non-being or negation is emphasised." In was Spinoza that said that all determination is negation.

This logical concept of negation was later adopted into the ontological scheme of the Vaiśeṣika and made into the new category of non-existence. The employment of this category in the Syncretist school has been very extensive. According to Athalye, the wonderful accuracy of the Indian syllogism, the processes of reasoning and analysis have been greatly facilitated by the recognition of abhāva (non-existence). The notion of non-existence is claimed to possess as much reality as its opposite. This is stated in the form of a pratiyogi and anuyogi relation, that is, every entity involves at the same time the conception of its counter-entity and *vice versa*.

There is distinct difference of opinion between the Naiyāvikas and *Vaiśeṣikas* as to the perceptibility of abhāva or non-existence. The former hold that it is an object of perception, while the latter that it is only an object of inference. The former go a step further and make non-existence consist of several kinds while properly speaking negation is simply non-existence in general. "All negation is pure and characterless" according to Athalye. In the Syncretist school, the conception of abhāva is employed in the sense in which the later Nyāya employs it, *i.e.* as consisting of many kinds and as being as many as there are conceivable counter-entities. This is one of the conception of the Nyāya-Vaiśeṣika system which has enabled it to develop a very subtle method of intellectual analysis."[43a]

It must, however, be said that in spite of the fact that *abhāva* has been of very great use for intellectual discussion, the Vaiśeṣikas have not laid out any procedure by which one can, step by step, arrive at the concept of *abhāva*. Neither the whole discussion nor the characteristics and types of *abhāva,* as given in the Nyāya-Vaiśeṣika systems of philosophy, are, therefore, of any use in mathematics. Nor again the *abhāva* of the Nyāya-Vaiśeṣika system of philosophy is in any way comparable to the zero in mathematics.

19.2.3. *The zero in Pāṇini's grammar;* [44]

As we enter into Pāṇini's field, we find every term well-defined in clear words, leaving no chance for any ambiguity. And once we master the definitions and the techniques, the grammar looks as simple as any work of scientific nature should be.

The zero in Pāṇini's grammar can be classified for the convenience and ease of understanding into two broad types. One is 'the express zero' which is laid down by explicit terms like *lopa, luk, ślu* or *lup*. The other is 'the implied zero', which is not laid down in explicit terms like *lopa, luk, ślu* and *lup,* but which, in ultimate analysis and descriptions, amounts to zero. Pāṇini uses different terms for 'the implied zero', like *ekādeśa, śeṣa* etc.

Sometimes Pāṇini intends the zero in a round about way by stating that a particular group of sounds is to be preserved (*Śiṣyate*), intending thereby the *lopa* of other sounds in that situation. We will now illustrate by examining the definitions and the situations in which the two types of zero are available in Pāṇini's grammar.

19.2.3.1. *The express zero*: The *sūtras* which state the express zero are two, viz. *adarśanam lopaḥ* , 1.1.60 and *pratyayasya luk-ślu lupaḥ,* 1.1.61. The first *sūtra,* 1.1.60 defines *lopa,* which etymologically (from √*lup* 'to disappear) means 'disappearance', and the second *sūtra,* 1.1.61 defines the zero by the terms *luk, ślu* and *lup.* The only difference between the two zeroes laid down by the two *sūtras* is that while the zero laid down by the term *lopa* is empowered with bringing about the morphological/phonological/accentual changes in the stem, the zero stated by the other three terms is not. But both the *sūtras* state the express zero.

The straight, literal meaning of the two *sūtras* is: if we find anywhere the disappearance (*a-darśana,* lit. non-appearance) of any entity which should have been or is stated to be there, we should understand that it is called *lopa* of the entity i.e. it is zeroed.[45] The root √*dṛś* (= 'to see') is to be taken here in a general sense of 'knowledge'[46]. According to Kāśikā, the word *adarśana* signifies the following meanings : *aśravaṇam* (non-hearing), *anuccāraṇam* (non-pronunciation), *anupalabdhiḥ* (non-availability), *abhāvaḥ* (non-existence) and *varṇavināśaḥ* (loss of sounds). All this means that the categories stated in the situations concerned are zeroed if the terms *lopa, luk, ślu* and *lup are* used with reference to them. It must be repeated here that for all practical purposes, all the four terms signify the same sense, viz. of zero. Thus, *lopa* = *luk* = *ślu* = *lup*.

Let us illustrate the working of this zero-technique by an example, in which the zero of the concerned category is brought about by the term *luk*. The example is provided by the pres. 3rd sing. form *atti* from √*ad* 'to eat' from second conjugation, which exhibits no *vikaraṇa*. The process is as follows:-

ad+śap+ti (*śap* is a *vikaraṇa*; *ti* is a *pratyaya*)

= *ad+a+ti* (*ś* and *p* = o)

Then by the *sūtra, adiprabhṛtibhyaḥ śapaḥ,* 2.4.72, the *śap* i.e. a is zeroed and we have the situation.

ad+o+ti

= *ad+ti*

= *at+ti*

= *atti.*

This whole process is to be read in the context of the one for the form *bhavati* (from √*bhū*, 1st conj.) in which there is no zero of the *vikaraṇa*. The process is: *bhū* + *sap+ti*

= *bhav* + *śap* + *ti*

= *bhav* + a + *ti*

= *bhavati*

19.2.3.2: *The implied zero* : The best example is provided by the compounds called *ekaśeṣa dvanda,* in which, as the name itself suggests, out of the two or many *prātipadika* - categories constituting its membership, one category remains with the result that the other/s is/are lost i.e. zeroed. Let us take the example of the *ekaśeṣa dvandva* compound *putrau,* meaning 'the son and the daughter.' What Pāṇini does is that he starts with two constituents,[47] *putra* and *duhitṛ* and then keeps the word *putra,* the other member *duhitṛ* in effect becoming zero. Thus we have, *putraḥ* + *duhitā* + au

= *putra* + *duhitṛ* + *au*

= *putra* + O + *au* (duhitṛ = O)

= *putra* + *au* which is

= *putrau*, which is the *dvandva* compound with the technique of *ekaśeṣa* zero. The *sūtra* which reduces *duhitṛ* to zero is *bhrātṛputrau svasṛduhitṛbhyām* , 1.2.68.

19.2.4. *The peculiarities of the Pāṇinian zero:*

The Pāṇinian technique of zero seems to be based on certain assumptions and brings out certain salient features.

The basic assumption of the Pāṇinian description of the Sanskrit language is that he starts his linguistic description with what **Sergiu Al-George** calls as "the richer comprehension"; that is to say, with a maximum or at least more expanded form; and then he goes on reducing or subtracting the unwanted elements to arrive at the minimum form available in the language. Thus in the example of *atti,* the model on which Pāṇini is working is of the form *bhavati,* which has a richer comprehension than *atti. bhavati* is conceived to be comprising of three-morpheme structure, viz. the *dhātu bhū,* the *vikaraṇa śap* and the *pratyaya ti.* And in order that the verbal form *atti* must also contain three morphemes on the pattern of *bhavati*, that Pāṇini assumes initially a three-morpheme structure for *atti* and then zeroes the undesired *vikaraṇa śap* which is not available in the spoken language. Thus, *atti* = *ad+śap+ti*

= $N + S_1 + S_2$ (N= Nucleus, S = suffix)

= $N + O + S_2$

= NS_2 which is the desired form.

If the above line of thinking is correct, the zero in Pāṇini's grammar is available by the process of substraction alone, and not by any other process. The structural approach combined with the process of subtraction alone gives the zero. Pāṇini zeroes the unwanted elements because they are not found in the language. Yet, he assumes originally the presence of such elements because he is very careful to see that the different entities or categories like *prātipadika, dhātu, pratyaya* etc. get their proper places in the formative process, though not in the final, usable form.

This leads us to discern yet another characteristic feature of Pāṇinian technique of linguistic description, viz. Pāṇinian description is based on positional analysis. It will be seen from the examples of *atti* and *bhavati* that the *suffix a* i.e. *śap* is first assumed in *atti* and then elided, since it is not found in the language. Such an assumption of a non-existing suffix in final analysis implies, first, a basic comparison between two types of formations from one group (here the group is of the verbal forms) and, secondly, the positional analysis of the formations to find out one-to-one correspondence between one form and the other. Thus the three morphemes in the form *atti* are positionally set against those in another form *bhavati* and a search for finding out the one-to-one correspondence between them is carried out by the formula, Form $= N + S_1 + S_2$. In this process, the *śap* is missing in *atti*; it is therefore, represented positionally by zero in the stage of the formative process itself. Thus,

ad + śap + ti

= *ad + O + ti*,

This type of analysis immediately gives out the one-to-one correspondence between *atti* and *bhavati*. The zero, which represents *śap*, has thus a reserved position in the whole analytical process.

Besides these references to positional analysis which can be drawn by implications, we have direct *sūtras* by *Pāṇini* which refer to the position of the different grammatical entities. The *sūtras* pratyayaḥ, 3.1.1. and *paraś ca*, 3.1.2. explicitly state the position of the *pratyayas* as posterior to the *prakṛti*. The *sūtra*, *ādyantau ṭakitau,* 1.1.46, refer to the initial and final position of the *āgamas*; *mid aco'ntyāt paraḥ,* 1.1.47 refers to position posterior to the last vowel in the base; The *sūtras, tasminn iti mirdiṣṭe pūrvasya,* 1.1.66 (prior position), *tasmādity uttarasya,* 1.1.67 (latter position), *alo'ntya sya,* 1.1.52 and *ṅic ca,* 1.1.53 (final position), *ādeḥ parasya,* 1.1.54 (initial of the posterior category) define clearly the positions of the suffexes to be applied. For the word, 'position', Pāṇini has used the word '*sthāna*' meaning 'place, position' etc.

The 'zero' in *atti* has, therefore, a 'positional value'; it indicates the 'position or place' of the suffix *śap.* It is to be noted that the 'zero' of *śap* is obtained here by subtraction, the addition being stated already. First it is added and then substracted.

19.3. Comparison of the philosophical systems with mathematics

We are now is a position to compare the concept of *śūnya, abhāva* and *lopa*, all in final analysis signifying zero, with that of zero in mathematics. Since all the number-systems in post-Vedic times are decimal number-systems, the characteristics which are available in them are also found in the Vedic number-system. And since the post-Vedic number-system is decimal, the Vedic system, as we have seen before, is decimal. Since again the post-Vedic number-system is based on the positional or place-notation, the Vedic number-system also must have been based on place-notation. If this is true, then the only philosophical system which offers a possibility of comparison with the mathematical system of the Vedas is of Pāṇini, in which, as we have seen before, the positional analysis is resoreted to. The Buddhistic system of philosophy expounding Śūnyavāda does not lay down or proceed on step-by-step positional analysis; it also does not lay down any formative process to arrive at the *śūnya*. The exposition of *abhāva* of the Naiyāvikas also does not spell out any process or positional considerations to arrive at the *abhāva*; rather, it seems to have the temporal implications at its root, since it considers *abhāva* with a reference to priority or posterity of the *kārya* and *kāraṇa*. Again, it does not state any formative process to arrive at *abhāva*. Moreover, unlike in mathematics, the concepts of *śūnya* in Buddhism and *abhāva* in Nyāya are not techniques employed to serve some other purpose, but principles; in mathematics the concept of zero is a technique employed to build up a number-system based on ranks. And it is only because of zero-technique that mathematics has been able to build up a number-system based on ranks. The only system, in which zero is employed as a technique is of Pāṇini. It is only with the Pāṇinian technique of *lopa* that the mathematical zero can be compared.[48] It is only Pāṇini who gives a regular, cosistent formative process to arrive at the zero; *cf.* his description of the forms *bhavati/atti* and of the *ekaśeṣa dvandva, putrau,* given before

19.3.1. *Two types of languages*

Before we proceed on to compare Vedic mathematical zero with the one in Pāṇini's grammar, we must take note of a point which is of utmost importance for our purposes. Veda is not a text-book on mathematics. As such, we will not find any direct references to mathematical theories. Whatever theories we have found out up till now are only by implication.

The second point to be noted is that Veda is composed in language and not in number-symbols. We will not find, therefore, any symbol for any number, much less for zero, in it. We will have to imagine the number-symbol for zero by implication and with the help of the language itself, that is to say, with the help of the number-words themselves which are used for numbers. At this stage, we can see that the Vedic language can be classified into two types; One, that language or words which describe and analyse the non-mathematical facts, like the praise of the gods etc. The facts may relate to any thing from simple speech to sacrifice, rituals, praises, songs or even philosophical considerations. We call this type of language as non-mathematical language. The second type of language which is used is full of words for mathematical numbers. It is these mathematical number-words that we have been discussing up till now. We may call this language as mathematical language. All the number-words and the words which indicate any mathematical operation like addition, subtraction etc. can be grouped under the category of mathematical language. All other non-mathematical words can be grouped under non-mathematical language. Thus in the passage, *dvādaśa pradhayaḥ, cakram ekam, trīṇi nabhyāni, ka u tac ciketa; tasmin sākam triśatā na śaṅkavaḥ arpitāḥ ṣaṣṭir na calācalāsaḥ,* R.V. 1. 164.48, the words *dvādaśa, ekam, trīṇi, triśatā, ṣaṣṭiḥ* are mathematical language, giving out mathematical information, while all others are non-mathematical language, giving out non-mathematical information. The mathematical words signifying numbers are, no doubt, linguistic entities, in the sense that all the rules of Sanskrit declension and conjugations are applicable to them. But over and above this, the information they give is purely mathematical. Hence they are mathematical entities also. It is for this reason that the two types of word-structures must be distinguished from each other, so far as their semantic import is concerned.

19.3.1.1. *aṅkānām vāmato gatiḥ :* That the Veda contains two types of languages — one, mathematical and the other, non-mathematical — is clear. Since both the types are languages, they are to be understood properly. Are they to be understood — or rather read — in the same way? Can we know the meaning of the mathematical language in the same way as we know the meaning of the non-mathematical language? To illustrate, shall we understand or read or write in symbols the number-word *dvādaśa* as 12 or 21 - the latter symbol follows the wording of the compound word-structure *dvā* (=2) and *daśa* (= 1; *das'a* is represented by 1 for which see below.) It is clear, therefore, that the two types of languages — mathematical and non-mathematical—cannot be understood or read in the same way, but require different methods for understanding. The non-mathematical language, say a word like *vajra—bhṛt*, is understood in the order in which the words are set therein. Thus, the word *vajra* being in the prior position is understood to give out the meaning first; then comes the word *bhṛt* which occupies the posterior position in that order. But it is not so in the case of the mathematical language, say a word *dvādaśa* in which traditionally *daśa* is understood first and then comes the turn of the word *dvā*. The two types of languages, though composed of the same Sanskrit phonemes, morphemes and words, require different ways of understanding. There will be a confussion if we understand them in the same way.

It is in order to avoid the confusion between the two that the Sanskrit mathematicians in Vedic times have spelled out a dictum for reading the two types of languages. The dictum is very famous in Sanskrit mathematics and runs as follows:- *aṅkānām vāmato gatiḥ*, which literally means 'the understanding (*gati* from √*gam* 'to go', also 'to understand') of the numbers (is to be done) in the reverse way (*vāmataḥ*)'. The phrase in *anuṣṭubh* metre seems to be — and is actually — a part of a verse in *anuṣṭubh* metre which runs as follows:-

aṅkeṣu śūnyavinyāsāt vṛddhiḥ syāt tu daśādhikā/

tansmāt jñeyā viśeṣeṇa aṅkānām vāmato gatiḥ .[49]

The word *aṅka* in the verse refers to the number-symbols for the numbers are to be written down or read or understood first in the order in which they are spoken and then the order is to be reversed. Thus, the word *dvā-daśa* is to be first understood in symbols as 2-1 and then the figures 2 and 1 are to be read in reverse position as 1-2 and the number will be written as 12.

In the system of writing Sanskrit which writes from left to right, the reverse order will be from right to left; that is, the number to the right, which is written last, will occupy the first position and then the numbers from right to left will occupy positions in that order. The word *dvā-dasa,* written as 2-1, will be then *daśa-dvā* and written as 12 which is the correct writing or understanding of the numbers.

The word *aṅka* may refer to the number-word also; the gen. *aṅkānā.m* will then mean *aṅkārthakaśabdānām*, 'the number-words conveying the meaning of numbers'. Yet the final result is the same.

The dictum of *aṅkānām vāmato gatiḥ,* though nowhere spelled out in any authoritative work on mathematics, Vedic or post-Vedic, seems to have been known and followed even in Vedic times also as the number-words used therein are to be reversed. The Persians and the Arabs, who have borrowed the number system from India seemed to have applied the Vedic technique of reversion of the order to all the words even to non-mathematical irrespective of the fact whether the words convey the mathematical meaning or not. Hence perhaps the whole system of their writing is the reverse of the Indian system, viz. from right to left. But interestingly enough, they do not reverse the numbers from right to left. They write the numbers from left to right itself after the fashion of the Indians.[50]

The rule *aṅkānām vāmato gatiḥ* is, therefore, of utmost importance in understanding the numbers. The rule actually seems to have been spelled out to understand the Vedic texts as principle of interpretation of Vedic language and seems to have been coined out of necessity. But, surprisingly enough, later post-Vedic writers adhered to this rule strictly and wrote the numbers purposefully and unnecessarily in the reverse fashion i.e. from right to left; *cf. Sūryasiddhānta* (9.5.3) (400 AD):

nava-vasu-sapta-aṣṭa-kha-nava-aśva

= 9 8 7 8 0 9 2, which is

= 2 9 0 8 7 8 9

or, *Pañcasiddhāntikā* (505 AD; 1.5.17);

śūnya — dvi — pañca — yama

= 0 2 5 2

= 2520.

Many example can be quoted from Sanskrit literature.

Not only this, but the later writers used symbolic words for number-words, (- perhaps for sacredness or secrecy? —) and complicated the easy understanding of mathematics. But we are not concerned with this at present. So far as Veda is concerned, the numbers are mentioned only by number-words and not symbolic words for numbers.

The word *vāma* is peculiar here. It signifies the sense of 'left, or reverse' in classical Sanskrit. It means 'beautiful' in the Vedic language and is always an adjective of *vasu* i.e. wealth. *Vāma* in the sense of 'reverse' indicates or refers to the direction opposite to the one in which one is proceeding. The west, for example, will be the *vāma* direction of the east, and vice-versa. The *devanāgarī* is written from left to right; the vāma of this direction will be from right to left. The 'above' is the *vāma* of 'below' and so on.

An important point requires to be noted. In solving the compound number-words and the compound non-number words, the same grammatical method can be or is resorted to. For example, the compound number-word *ekādaśa* is dissolved as *ekaḥ ca daśa ca*; similarly, the *dvandva* compound *indrāgnī*, which consists of non-number words, can also be, or is, solved in the same way, viz. *indraḥ ca agniḥ ca*. What is the point, therefore, in explicitly stating the principle *aṅkānām vāmato gatiḥ* in the case of the number-words alone? In the case of both the types of compounds, numerical and non-numerical, the positions of the constituents remains the same in the stage of dissolution; the word *eka* occupies the first position in the compound (*samāsa*) as well as in its dissolution (*vigraha-vākya*); and there seems to be apparently no significant purpose is stating the above dictum of *aṅkānām vāmato gatiḥ*. The only purpose that seems to be implied behind the above dictum is that the principle aims at laying down the order of the number-symbols while writing them down. This leads us to the next condition that writing of the numbers at least in symbols was prevalent in the Vedic and the immediately succeeding post-Vedic times. Without such as presumption, the dictum *aṅkānām vāmato gatiḥ* seems to be meaningless. Though the principle seems to have been enunciated later, it was applicable in Vedic times also.

The dictum *aṅkānām vāmato gatiḥ* is, to put it in the grammatical language, a kind of *paribhāṣā* which helps to read or understand the mathematical language, just as the *paribhāṣās* in grammar help to understand the implications of Pāṇinian *sūtras* in the case of any confusion. The only difference in the above mathematical *paribhāṣa* is that while the latter grammatical *paribhāṣās* are collected together and recorded by the grammarian Nāgoji Bhaṭṭa in his book called *paribhāṣenduśekhara,* the former mathematical *paribhāṣās* are nowhere collected together and recorded. They have been only handed down orally through *guru-śiṣya-paramparā.*

We thus can see that since there are two different languages in the Veda, there are two different ways of representing them in different symbols—the *varṇa* - symbols for the non-mathematical language and the *aṅka* symbol for the mathematical language. Since again there are two types of symbols for the two languages, there are two types of ways of reading or understanding them, viz. from left-to-right for the non-mathematical language and right-to-left for the mathematical language. The whole method is, therefore, perfectly scientific and logical.

19.3.2. *The Sanskrit Word-Structures*

Besides the two important points which have been noted above, viz. existence of two types of words or languages in the Veda and the principle *aṅkānaṃ vāmato gatiḥ* spelled out for the interpretation of the mathematical language in the Veda, we have to take note of the third important point which relates to the representation of the Sanskrit word-structures in terms of symbols. The representation of the words in symbols is based on the Pāṇinian analysis of the Sanskrit language.[51]

If we turn to Pāṇini for the analysis, description and explanation of the Sanskrit word-structure, we find that he starts with the implied assumption that in Sanskrit, the *pratyaya,* referring primarily to the terminations i.e. closing morphemes is a compulsory category. No word can be used in the language without applying the *pratyaya,* which includes both the declensional (i.e. *sup*) and the conjugational (i.e. *tiṅ*) closing morphemes. The *prakṛti,* referring to both the nominal as well as the verbal base and the *pratyaya* are thus attached with each other enternally. We cannot use or even imagine one without the other. This Pāṇinian assumption underlying his analysis and description of the Sanskrit language is made explicity clear by Patañjali in unequivocal terms. *cf.* Patañjali, the commentator on Pāṇini's *Aṣṭādhāyi* on *the Pāṇinian* sūtra, 1.2.45: *pratyayena* nityasaṁbandhāt. nityasaṁbandhāv *etāv arthau prakṛtiḥ pratyaya iti. pratyayena nitya-saṁbandhāt kevalasya prayogo na bhaviṣyati.*

19.3.2.1. *The non-compound word-structures* : This explanation by Patañjali of Pāṇini's underlying assumption helps us to represent the Sanskrit word-structure in convenient symbols. Let *prakṛti,* which is the nominal and/or verbal base, be represented by the symbol N signifying Nucleus; Let *pratyaya,* which refers to the non-closing morphemes of the *taddhita, kṛdanta* and feminine spheres as well as to the closing morphemes (viz. *sup* and *tiṅ*) i.e. terminations, be represented by the symbol S signifying in general the suffix. Let F represent the final form which is used or usable in the language. Any non-compound word-structure in Sanskrit can now be represented in the formula, F = N.S. The structures, say, *rāmasya* or *caturbhiḥ* can both be represented by the formula, F = N.S. Thus,

rāmasya = *rāma+sya*

= F = N.S.

Also, *caturbhiḥ* = *catur* + *bhiḥ*

= F = N.S.

There are many types of structures in Sanskrit, simple and complex, comprising suffixes numbering from zero to infinity.

19.3.3. *The compound word-structure*

Coming to the description of the compounded word-structures, we find that Pāṇini describes them in the *sūtra, saha supā,* 2.1.3.

The *sūtra* states that any declined form used in the language (called *sup-anta* in Pāṇini's terminology) ending in the nominal declensional terminations (called the *sup-pratyayas* in Pāṇini's terminology and enumerated in the *sūtra*, 4.1.2) can be compounded or juxta-posed (*cf.* the Pāṇinian term for compound, viz. *samāsa* which is derived from *sam* + √*ās*, 'to be or put together) with any other declined form ending in nominal declensional terminations. Thus, the two *sup-anta* formations *rājñaḥ and puruṣaḥ* can be compounded together. The process is as follows:

rājñaḥ + *puruṣaḥ*.

Now, the word *rājñaḥ* can be split up as *rājan* (which is the *prakṛti*) and *as* (which is the pratyaya); and the word *puruṣaḥ*, as *puruṣa* (the *prakṛti*) and *-s* (the *pratyaya*); and we have,

(*rājan+as*) + (*puruṣai+s*) = (N+S)+(N+S)

Then according to the *sūtra*, *supo dhātuprātipadikayoḥ*, 2.4.71, the suffixes *-as* (of *rājan*) and *s* (of *puruṣa*) are zeroed; and we have the following picture, (*rājan*+O) + (*puruṣa*+O) = (N+O)+(N+O) = *rāja-puruṣa* = N.N., which is the compound formation. The formula in terms of the symbols F, N and S for a compounded formation is Sanskrit, therefore, will be:

F = N+N = N.N, to which the declensional suffixes viz. *sup-pratyayas* are again applied for using it in the language.

In terms of symbols, therefore, the two types of Sanskrit formations, viz. the non-compounded and the compounded, can be defined as: the formation which has only one N is non-compounded and the one which has more than one N is a compounded formation. Thus,

the *a-samāsa* F = $N+S_1+S_2+S_3$ - - Sn — S∞, and the *samāsa* F = (N+N) + $N+S_1+S_2+S_3$ - - Sn — S∞,

19.3.4: *The structure of the number-words*

The same rules and suffixes which are applied to get the *non-*number words as given above are applicable *in toto* to the number-words also, since the number-words, being nominal bases or *prātipadikas*, are on par with non-number words. As such, all the number-words used is the language, Vedic and classical, can be safely represented by the same symbols and formula as is used in the case of the non-number words. Thus,

F = *ekaḥ* = eka +s = N+S;

or, F = *dvau* = *dvi+au* = N+S;

or, F = *pañcabhiḥ* = *pañca + bhiḥ* = N+S;

or, F = *saptabhyaḥ* = *sapta+bhyas* = N+S;

or, F = *navabhiḥ* = *nava+bhiḥ* = N+S;

or, F = *daśasu* = *daśa + su* = *N+S* ; and so on.

As all the above structures of the number-words contain one, single N, they are grammatically, non-compound word-structure, and also consequently mathematically non-compound numbers.

If we try to examine and represent the number-words after *daśa*, the picture that we get is like the following: Let us take the example of the number-word *ekādaśa*. It is dissolved in the Veda itself as *ekā ca... daśa ca,* A.V. 5.15.1; the feminine gender used is the text is not relevant to the present discussion. We can even take the masculine form as *ekaḥ* instead of the fem. *ekā* in the text for the purposes of displaying the grammatical dissolution. Thus, we can safely dissolve the word *ekādaśa* in the masc. as *ekaḥ ca daśa ca*. And after the pattern of the Pāṇinian process, we have,

F = *ekādaśa*

= *ekaḥ* + *daśa* = (N+S) + (N+S)

= *eka* + *daśa* (with *s*=O) = (N) + (N)

= N + N

= N.N, to which the suffixes of the nominal declensions viz. the *sup-pratyayas* (listed in the *sūtra*, 4.1.2) can be applied for using it in the language, since in the form in which it is available, it is a word, or more correctly a *prātipadika*.

All other number-words after *daśa* can be represented symbolically in the same way. Thus,

F = *dvādaśa* - *dvi+daśa* = N_1+N_2

or, F = *trayodaśa* = *tri+daśa* = N_1+N_2

or, F = *ekaviṁśati* = *eka+viṁśati* = N_1+N_2 and so on.

Since, as will be clear from the above symbolic representation, all the structures of the number-words from *ekādaśa* onwards contain two N, they are grammatically compound word-structures, and also consequently, mathematically compound numbers.

19.4. The type of compound

That the number-words from *ekādaśa* onwards are compound word-structure is clear beyond doubt. But what type of compounds are they?

As we know, there are four main types of compounds in Sanskrit according to Pāṇini. They are *avyayībhāva*, *tatpuruṣa*, *dvandva* and *bahuvrīhi*. The *avyayībhāva*- compound is defined in the *sūtras*, 2.1.5 and 2.1.6; the *tatpuruṣa* is defined in 2.1.22; the *dvandva* is defined in 2.2.29 and the *bahuvrīhi* is defined in 2.2.23 and 2.2.24. If we apply the criteria for being a particular type of compound given in the Pāṇinian *sūtras*, we find that the above compound number word-structures satisfy only the criterion given for the *dvandva* compound in the *sūtra*, 2.2.29, *cārthe dvandvaḥ*. What the *sūtra* means that if two *sup-antas* i.e. nominal bases are joined together by the semantic relation of '*ca*' i.e. 'and', the compound they form is called the *dvandva* compound. We have seen above that the two nominal bases *eka* and *daśa* (and *dvi, tri* etc. and *daśa* for the matter) are joined together by the semantic relation of '*ca*' i.e. 'and'; thus, *ekādaśa* means *eka* and *das'a* (i.e. one and ten);

hence the compound that all the above number word-structures form is a *dvandva* compound. The word-structures from *ekādaśa* to *navadaśa* (11-19), *ekavimiśati* to *navaviṃśati* (21-29), *ekatriṃśat* to *ekonacatvāriṃśat* (31-39), *ekacatvāriṃśat* to *ekonapañcāśat* (41-49), *ekapañcāśat* to ekonaṣaṣṭi (51-59), *ekaṣaṣṭi* to *ekonasaptati* (61-69), *ekasaptati* to *ekona-aśīti* (71-79), *eka-aśīti* to *ekona-navati* (81-89) and finally, *ekanavati* to *nava-navati* (91-99) are all *dvandva* compound word-structures and can safely be represented symbolically by the formula given above, viz. $N_1+N_2 = F$. These are in all 81 *dvandva* compound word-structures. The remaining 19 number-words out of one hundred, from *eka* to *nava* (1-9) and *daśa* (10), *viṁśati* (20), *triṁśat* (30), *catvāriṁśat* (40), *pañcāśat* (50), *ṣaṣṭi* (60), *saptati* (70), *aśīti* (80), *navati* (90) and *śatam* (100), do not seem to be compounds at all. The question of the numbers *śatam* and above does not arise at all here since these numbers are not expressed in terms of compound number-words; they are expressed by phrases. The only numbers above *śatam* which are expressed by non-compounded single number-words are *sahasram*, *ayutam, niyuta, prayuta, arbuda, nyarbuda, samudra, madhya, anta,* and *parārdha* which we have already discussed above; *cf.* VS. 17.2. These number-words also again are not compound word-structures, except the words *a-yuta, ni-yuta, pra-yuta, ny-arbuda* and *parārdha*. The peculiarity of these latter compounds is that none of the members comprising them is a number-word.

If we closely examine these facts about, which number-words are compounded word-structures containing two number-words and which are not so, we find that the number-words which serve as the radix or basis for either the next series (as in the case of *daśa*, *viṁśati, triṁśat*, *catvāriṁśat, pañcāśat, ṣaṣṭi, saptati, aśīti* and *navati* which are bases for the series next to them) or next ranks (as in the case of *śata, sahasra* etc. which are the basis for the ranks after them) are expressed in simple, non-compounded word-structures and not in complex, compounded word-structures. And those number-words above *daśa* which are not radix words are expressed in compound word-structures containing two members, N_1 and N_2, one of which, preferably the last or posterior or second, is the radix word. Why? There are two possible explanations of this phenomenon: (1) either the radix words are or were basically and originally simple, non-compounded words, or (ii) they were originally complex, compounded word-structures, with one member, preferably the non-radix number-word occupying the first or prior position, lost.

But this line of thinking raises many problems which must be answered.

19.5. The radix and non-radix number-words:-

We have divided the Vedic number-words above into two main types, viz. the number-words which serve as the radix, and those which are not radix number-words. What we mean by the radix number-words is that these words first linguistically and then mathematically are to be read or grouped or go with the immediately succeeding series and not with the preceding ones. Thus, linguistically and mathematically the word *daśa* goes with the succeeding series of numbers from *ekādaśa* to *navadaśa* and not with the preceding series viz. from *eka* to *nava*. So also with all the other radix number-words *viṁśati*, *triṁśat* etc. In other words, the word *daśa*, for example, is not the last one for the series from *eka* to *nava*, but it is the first, or at the head, of the series from *ekādaśa* to *navadaśa*. The obvious reason for such a grouping, surprisingly enough, is originally purely linguistic and not mathematical, though later on it is true mathematically also. If this line of thinking is correct, the radix words *daśa* etc. must be linguistically compared or explained with reference to the number-words in the succeeding series and not with those in the preceding series. We have to compare them linguistically and not mathematically; the reasons are as follows:-

1. The Veda is a language and contains only words, and not symbols, for the numbers;

2. We have no definite proof that the Vedic number-words were ever translated into number-symbols in Vedic times themselves.

We have, therefore, no alternative other than linguistic for the comparison of the radix words with non-radix words.

19.6. The birth of Zero

Let us now start comparing the radix words and the non-radix words linguistically with the help of the symbols to signify the nominal bases in the compounds. The radix *daśa* is taken as an example. The same arguments, procedure and representation as are used is the case of *daśa* also hold good in the case of all other radix-words, viz. *vīṁśati, triṁṣat* etc. The symbol N used before would have done as a general symbol for nominal bases. But since we have here different nominal bases and have to distinguish them from one another, we prefer to use different symbols for the different nominal bases. Since *daśa* is the same nominal base, we have kept the symbol N for *daśa* only. The nominal symbols used here are:

a for *eka;*	*q* for *ṣaṭ* > *ṣ* o
b for *dvi;*	*x* for *sapta*
c for *tri*	*y* for *aṣṭa*
d for *catur*	*z* for *nava*
p for *pañca*.	and N for *daśa*.

We now represent the series from *ekādaśa* to *navadaśa* symbotically as follows:

ekādaśa = *eka* + *daśa* = a. N.

dvādaśa = *dvi* + *daśa* = *b*. N.

trayodaśa = *tri* + *daśa* = *c*. N.

caturdaśa = *catur* + *daśa* = *d*. N.

pañca das'a = *pañca* + *daśa* = *p*. N.

ṣodaśa = *ṣaṭ* + *daśa* = *q*. N.

saptadaśa = *sapta* + *daśa* = *x*. N.

aṣṭādaśa = *aṣṭa*+ *daśa* = *y*. N., and finally

navadaśa = *nava* + *daśa* = *z*. N. (Note that *nava-daśa* is more natural to the Vedic technique of building up the number structure than *ekonavimśati* which seems later.)

Now, since the radix word *daśa* is in the beginning of this series, its word-structure is also to be read and compared with the structures of all the members in the series beginning with *daśa*.

Since all the non-radix number-words in the series from ekādaśa to *navadaśa* contain two N and hence are compounded word-structures linguistically, the radix-word *daśa* must also contain two N, must consequently be taken as a compound word-structure and must, therefore, be represented as N_1+N_2, with N_2 representing *daśa* since the word *daśa,* as a radix, occupies the posterior or second position in the above series. The symbolic representation of *daśa* as a compound word-structure must then be:

daśa = $N_1 + N_2$

= N1 + *daśa*

But we have no N_1 here; we, therefore, represent N_1 as a vacant place.

daśa = vacant place + *daśa*.

If now we substitute the symbol of zero, viz. O for the vacant place, we have the formula for *daśa* as:

daśa = O + *daśa*.

Instead of comparing all the number word-structures from *ekādaśa* to *navadaśa*, even a single structure like *ekādaśa* will do and we have the following:

ekādaśa = a + N = aN, and

daśa = vacant place + N

And in the vacant place we put the modern symbol for zero, viz. O and we have,

daśa = O. N; and then by the dictum, *aṅkānām vāmato gatiḥ*, we get *daśa* = O.N = N.O. *daśa* thus assumes a two-digit or two-symbol form.

The same type of representation in symbols can also be done for the radix-words *viṁśati, triṁśat* etc. as

ekaviṁśati = *eka* + viṁśati = a + N = a. N and

viṁśati = Vacant place + *viṁśati* N = O.N. = N.O.

This formula, viz. O + N will bring the structure of *daśa* in line with the other structures of the series from *ekādaśa* to *navadaśa* with which the word *daśa* is read and of which he is at the head or beginning.

We can arrive at the same result with the help of the technique used by Pāṇini while describing the *ekaśeṣa dvandva* compound *putrau* in which the other member viz. *duhitṛ* is zeroed. And as we have seen above, all the non-radix number-words are nothing but the *dvandva* compounds. And the Pāṇinian technique applied in the case of the *dvandva* compounds in general can be safely applied in the present case also.

We repeat the technical process given by Pāṇini in the case of *putrau*.

putrau = *putraḥ* + *duhitā* + *au*

= *putra* + *duhitṛ* + *au* (suffixes = O; *cf.* 2.4.71)

= *putra* + *O* + *au* (*putra* remains, resulting in the zero of *duhitṛ*; *cf.* 1.2.68

= *putrau*.

In symbols,

$N_1 + N_2 + S$

$= N_1 + O + S$

$= N_1S.$

Similarly, just as Pāṇini has assumed a second member viz. *duhitṛ* for the sake of symmetry of linguistic description, we can also assume a suitable mathematical number-word viz. say, *eka* or *dvi* or any other number-word as the other member besides *daśa* and elide it. Thus,

daśa = *eka* + *daśa*

= O + *daśa*; which is a two-digit representation of *daśa*.

Now by following the dictum of *aṅkānām vāmato gatiḥ,* we can re-write the series from *daśa* to *nava-daśa* in terms of symbols as:

daśa = O.N. = N.O.

ekā-daśa = a.N = N.a

dvā-daśa = b.N = N.b.

trayo-daśa c.N = N.c.

catur-daśa = d.N = N.d.

pañca-daśa = p.N - N.p.

ṣoḍaśa = q.N = N.q.

sapta-daśa = x.N = N.x

aṣṭa-daśa = y.N = N.y.

nava-daśa = z.N = N.Z.

The N which occupies the second i.e. posterior position in column 1 comes to take up the prior position is column No-2; and *vice-versa,* the zero, a, *b, c, d, p, q, x, y* and *z* which occupy the prior position in column No.1 are reduced to the posterior position in column No.2. The right-hand positions represent the digital places; and the left-hand positions represent the decimal place. Thus the word *daśa* comes to contain two symbols for one concept. So also all the two-digit symbols, and for that matter all the multi-digit symbols in the following series based on different ranks come to signify single concept of the respective numbers, although they themselves exhibit the appearance of having more than one symbols and hence numbers, which is not true. The truth is that what the many symbols in the multi-symbol appearance of the higher numbers like 12344 or 15625 etc. indicate is not the numbers but the places or ranks. The left-hand side word *daśa*, which looks originally as non-compounded one now becomes a compound, or rather *ekaśeṣa-dvandva-* compound word on the right-hand side. The assumption of a second member, besides *daśa*, is motivated by two main considerations: (i) to portrait *daśa* as an *ekaśeṣa dvandva* compound, and (ii) to achieve symmetry in linguistic and mathematical description of the forms and numbers, since *daśa* as the radix number-word goes with the succeeding series from *ekādaśa* to navadaśa and not with the preceding series from *eka* to *nava*. Also, to portrait *daśa* as composed of two elements is necessitated by the fact that the number *daśa* is represented as composed of *two,* and not one, mathematical symbols as 1 with 0 i.e. 10. in all the written mathematical symbols without any exceptions in all the ancient manuscripts and inscriptions. The idea of the two-digit notation for 10 cannot come from void but must have some tradition behind it - mathematical or linguistic. And this tradition seems to be only the Vedic tradition alone in the form of language, which alone can guide us to arrive at the notation of *daśa* in two-digit symbol. The representation in two symbols, or as a compound number, of the number-word *daśa* has up till now not been explained satisfactorily by any book on mathematics. And hence this present attempt to explain it. Only the Veda provides an answer to this, and the answer is obtained by employing the Pāṇinian technique of linguistic description, applied to mathematical number-words from Veda. This seems to be the only explanation.

There is yet another difficulty. That by projecting *daśa* as an ekaṣeṣa *dvandva* compound we can explain the two-digit representation quite logically is clear. But, the *ekaśeṣa dvandva* compound *putrau* cannot be compared *in toto* with the *ekaśeṣa dvandva* compound *daśa*. There are many points of dissimilarrity between the two. First, in the case of the *ekaśeṣa* dvandva, *putrau*, it is the second member viz. *duhitṛ* which amounts to zero; in the present case of *daśa*, it is the first member viz. *eka* which amounts to zero. The result is that in the case of *putrau*, what we get as result in the form of a formula is N.O. while is the case of *daśa*, the result that we get is O.N. Though these two results look apparently identical, they are in fact not so, because the language may tolerate the identity of the two structures for *putrau*, viz. O.N., N.O., but mathematics cannot tolerate the identity of the two structures for *daśa* viz. O.N. and N.O., or more specifically in actual modern number-symbols, between 01 and 10. The obvious reason is that while in language, and especially in *ekaśeṣa dvandva* compound, the two words viz. *putra* and *duhitṛ* can occupy any position, first or second, without harming the meaning and the final structure, in mathematics, position of the number is very important; the symbols cannot interchange their positions without harming the value of the numbers; 01 and 10 donot convey in mathematics the sense of the same values. The consideration of position or positional value is very rigid is mathematics, while it is not so much rigid is language, though the Sanskrit language and its grammar by Pāṇini do consider positions of the words in certain cases as important.

Secondly, there is the difficulty of equating our present formula for *daśa* as O.N. with the modern symbol for *daśa* viz. 10. The formula that we get in the case of *daśa*, as we have seen above, is O.N. The zero here stands for the vacant place of the word *eka* and N stands for the word *daśa*. If now we translate the formula in modern number-symbols, the number-symbol for the word *daśa* comes out to be 10; and the formula O.N. takes the numerical form as 010 which is untrue since the formula gives us a three-digit number for *daśa*, with two zeroes one preceding and the other succeeding; we write the modern symbol for *daśa* as only 10, i.e. as a two-digit figure. What actually we expect the formula to give out is only 10, without any additional number-figure.

Thirdly, if we write the equation as ON = 01, deleting the zero succeeding the number-symbol 1 and then by the axiom of *aṅkānām vāmato gatiḥ,* reverse the positions of 0 and 1 and rewrite the equation O.N as N.O. = 10, we do get *daśa* = N O = 10 (*cf. aṅkānām vāmato gatiḥ*). But in this case, the N will stand only for the symbol 1, which, as is clear, goes against our original position in which we have substituted N for *daśa* and not for 1. And what we want is that the whole two-digit symbol 10 should represent *daśa*.

The last alternative leads us nearer to the solution why *daśa* is represented as a two-digit number. But the number-symbol 1 which stands for *daśa* is a problem.

There are the three alternatives in the case of assuming or justifying the symbol 1 for *daśa*: (i) Either we should write *daśa* or N in single-digit number-symbol. But this goes against all traditional practices, eastern and western.

(ii) or, we should accept the three-digit symbol viz. 010 for *daśa*, which is also against the age-old practice.

(iii) or, we should equate *eka* with zero and *daśa* with 1 in the procedure of dissolving the compound *ekādaśa* as an *ekaśeṣa-dvandva* compound. But this is also not possible because in the present conventions of the number-words, *eka* is numerically never equal to zero and *daśa* is never numerically equal to 1.

What is the way out of this symbolically technical difficulty? What do the symbols 1 and 0 in the symbol for *daśa*, 10, stand for?

The answer seems to be: that we consider the symbol N in ON or NO, or 1 in 01 or 10 as signifying numbers itself seems to be a wrong position. What the symbol 1 in 10 seems to designate is not the number 1 in the single-digit series 0,1,2,3 etc. but the 'position' of the succeeding series viz. from 11-19, which is the 'first' of all the two-digit series after the single-digit series from 1-9. The figure 1 in 10 is not one from the unit-figures, but indicates the number of the two-digit series. And the symbol 10 gives out the following meaning: 1 represents the 'first' number of the two-digit series that are going to follow hereafter; and 0 represents that there is no unit-number in the 'first number' of the succeeding two-digit series; and according all the succeeding two-digit numbers can be interpreted. Thus, 15 will mean 'the fifth unit-number in the first two-digit series' and so on. As the number of the compound series increases, the number representing the number of the series moves to the left. Thus in the number 5, there is no question of left- and-right-hand movement. But in 54, the figure 5 moves to the left representing the 'fifth' two-digit series; in 543, 5 stands for the 'fifth' three-digit series and so on. The radix words, therefore, are used to signify the number of the series and should not be confounded with the units 0,2,3, etc. They function as a sort of 'command' commanding the start of new series. And the new series cannot be represented without using two-digit symbols - one indicating the number of the series and the other the number of the units in the series. The first symbols, viz. 1,2,3 ... in, say, 10, 20, 30 act as an indicator of the number of the two-digit series. While committing the concept of the series to writing, these symbols occupy the position prior to the other number-symbols. Since writing requires space and since we cannot write the two-digit symbols is one place - one over the above, we have to write them in ordered position. Position is the characteristic of space. The second symbol of zero in the above numbers is the number proper. In a symbol for eleven, viz. 11, for example, though the two symbols are identical, what the first 1 signifies is the number of the series and what the second one symbolises is the number proper. In spite of their identity, their functions are totally different from each other's. The phrases, therefore, like *ekādaśa*, *ekaviṁśati* etc. and the corresponding number-symbols 11, 21 etc. really mean or indicate 'the first number in the series of *daśa* which is the first series; or in the series of *viṁśati* which is the second series' etc. Hence *daśa* can be represented safely by the symbol 1, vesting it with different function from that of the other identical symbol 1. There, therefore, is no harm if we equate N=*daśa* = 1 in spite of the two-digit representation of *daśa*. Since *daśa* is the beginning of the next series and contains no digit, the vacant place for the digit is represented by zero. Hence *daśa* can be safely represented by the two-digit symbol 10.

The consequent series from 10 to 99 will then safely be represented by two-digit symbols. The subsequent numbers from 100 to 999 can also conveniently be represented by the three-digit symbols like 100 for hundred, 101 for one-hundred-and-one so on. It must be remembered, however, that in the present system of place-notation, the figures or symbols to the left indicate the higher orders or ranks and as we go from left to right (in the system of writing from left to right), we descend to lower orders or ranks. The right-most figure will indicate the lowest rank of the single-digital series from 0 to 9.

In terms of the ranks and series, what a number-symbol like 1 2 3 4 will really mean is; 'the number 4 in the third series of 'tens' which is in turn in the second series of 'hundreds' which again is the first series or rank of 'thousand'. The number can also be variously read from right to left as; (i) the number 4 is in the 123rd series of 'ten'; or (ii) is in the 3rd series of 'ten's which are in turn in the 12th series of 'the hundreds' and so on. Thus, except the right-most number-symbol 4, all other symbols indicate only the number of the series in ascending order from 'ten' onwards, in which the position of 4 is determined; they do not indicate the actual number, but indicate only the position. It seems that only the single-digit numbers alone are real basic numbers which are not derived mathematically from any other numbers. All other numbers are derived numbers, derived on the basis of positions or ranks or series - whatever one may call that. In the technical terminology of the Sāṅkhya - philosophy, the numbers from 0 to 9 are the *prakṛtis* and all others are the *vikṛtis*.

Coming to the number-word *daśa* and its number-symbol 10, what the figure 1 on the left signifies is 'the number of the series; viz. the first series after the single-digit series' and not 'the number proper'; and what the symbol 0 signifies is 'the empty, un-occupied space' since the subsequent two-digit series which follows does not contain any number in its digital place.

Just as in Pāṇini's grammar, the zero substitutioned for *śap* in the form *atti* has its non-zero counter-part in the *śap* of the form *bhav-a-ti* (*a=śap*), similarly the right-hand 0 in the symbol 10 has its non-zero counter-parts in the symbols 11, 12, 13 etc. The counter-parts are the right-hand symbols 1, 2, 3, etc.

Though none of the Vedic texts taken here for study provides any direct evidence and explicit reference to the effect that the number-symbols to the left indicate 'the places' and not 'the numbers', it can be known from post-Vedic works on mathematics. Āryabhaṭa. I (C.476 AD) writes that the values of the places to the left hand increase ten times that of those to the right-hand in a system of writing from left to right; *Āryabhaṭīya*, ch. on *Gaṇita*, verse 2: *sthānāt sthānam daśaguṇam syāt.* Bhāskarcārya clearly states that the numbers to the left stand for 'placeś; *cf. Līlāvatī*, verse 12: *iti daśaguṇottaram saṁjñāḥ saṁkhyāyāḥ sthānānām kṛtāḥ pūraiḥ.* The remark '*pūrvaiḥ kṛtāḥ*' (= 'defined by the predecessors in the field') definitely suggests or implies that there is a long pre-Bhāskarācārya tradition which was following this dictum; and this tradition seems to be right from the Vedas.

The concept of 'the zero or *śūnya*' definitely originated in the Vedas themselves and is not a later, post-vedic invention. The invention or existence of *śūnya* as a substitute for a vacant, un-occupied place or position goes back to as old as the Vedic times itself and can be convincingly proved with the help of a suitable technique, either Vedic or Pāṇinian.[52] And it must be said that it is from and in Vedic mathematics that the zero is born. And it is from the Vedas that all other civilizations have borrowed the concept and device of positional notation of numbers and consequently of zero.

An important point requires to be noted here. Taking into consideration the logic and the procedure behind the concept of *Śūnya,* it is unthinkable that writing was unknown to the Vedic civilisation, though, it must be admitted that, there is no direct evidence and reference to the art of writing.

The concept of zero, which is based on the principle of positional notation of numbers, the explicit enunciation and mention of numbers from *daśa* to *parārdha* (10^{12}) in VS.172, the mention of very high numbers like 3339, 6000, 8000, 12066, 30000, 50000, 60000, 60099 and finally 100,000 in as old a Vedic text as the Ṛgveda (which is the oldest literature of the world) and the knowledge of mathematical operations of addition, subtraction, multiplication, division, squares, and the expansion of different series like arithmetic and geometric progression—all these facts which are noted before point to only one conclusion, viz. that this knowledge of mathematics and the play of numbers cannot be said to exist without the knowledge of writing numbers. It is impossible to imagine that the Vedic people carried on all the mathematical calculations with big numbers without writing. The same conclusion seems to be inevitable in the case of the Vedic interpretational principle, *aṅkānām vāmato gatiḥ*, for which see before. Though the principle is enunciated later in post-Vedic times, it must have been resorted to in the Vedic times also for the interpretation of the mathematical number-words and phrases of the Veda. It is only by assuming this principle that the numbers from 10 onwards can be written down in symbols in the way in which they have been written down from olden times and in all the mathematical systems of the world. True, that there is no direct, documentary evidence to show that writing was known to the Vedas. Yet, direct, documentary evidence, or what is called as the *pratyakṣa pramāṇa* is not the only authority. We can draw conclusions even on the basis of circumstantial evidence or *anumānapramāṇa* also.

From all the discussion above it will be seen that the discovery of the insertion of zero to indicate a vacant place in numbers is possible only when we first analyse and study the Vedic number-words linguistically or grammatically; and that too through Pāṇinian eyes. Secondly, even after the symbolic representation of the number-words based on the Pāṇinian analysis, the picture that we get, viz. O.N. (or, in figures 01) is exactly oppositte of N.O. (or in figures 10) which is our present practice. At this stage the mathematical *paribhāṣa*, viz. *aṇkānām vāmato gatiḥ* must be taken into consideration. It is also interesting to note that it is only the Pāṇinian technique of linguistic analysis which helps us to arrive at zero, though Pāṇinian technique is later than Veda.

To explain all the above arguments in the form of linguistic factorisation, we arrange the number-words from *daśa* to *navadaśa* in the following serial order:

daśa-ekādaśa-dvādaśa-trayodaśa-caturdaśa-pañcadaśa-ṣoḍaśa–saptādaśa-aṣṭādaśa-navadaśa

We then take out the common factor-word *daśa* and we have, = dasa {*eka-dvi-tri-catur-pañca-ṣaṭ-sapta-aṣṭa-nava*}

Since the word *daśa*, as we have said above, signifies the number of the series, and since this is the first two-digit series, succeeding after the one-digit unit-series from *eka* to *nava*, we represent *daśa* with the symbol1, which is the first number of the whole number-series; we also represent the number-words *eka, dvi, tri* etc. by their symbols; and we have,

1 {0-1-2-3-4-5-6-7-8-9}

We then solve the brackets and we have, 10-11-12-13-14-15-16-17-18-19.

Thus, we have the exact symbol-replica of the number-words from *daśa* to *navadaśa*.

It is to be remembered that the whole process of factorisation is on linguistic and positional symbolic level and not on mathematical level.

We can thus have the infinite number of series in following:

(A) The first single-digit series : (0, 1, 2, 3, 4, 5, 6, 7, 8, 9)

(B) The first two-digist i.e. *daśa* - series: 1 (0, 1, 2, 3, 4, 5, 6, 7, 8, 9)
The second two-digist i.e. *viṁśati*-series:
2 (0, 1, 2, 3, 4, 5, 6, 7, 8, 9)

(C) The ninth two-digist i.e. *navati*-series
9 (0, 1, 2, 3, 4, 5, 6, 7, 8, 9)

(D) The tenth two-digit or the first three-digist i.e. the *śatam*-series
10 (0, 1, 2, 3, 4, 5, 6, 7, 8, 9)

(E) The hundredth two-digit or the tenth three-digit or the first four-digit i.e. the *sahasra* -series:
100 (0, 1, 2, 3, 4, 5, 6, 7, 8, 9)

(F) The *n*-th, *m*-digit series :
n(0, 1, 2, 3, 4, 5, 6, 7, 8, 9) and so on.

We can thus go on expanding or erecting the number-series until we come to the end of both space and time. Even then the series will still have scope of expansion. The whole process and the infinite nature of the number-expansion reminds us of the description of greatness of Lord Śiva given in the *śiva-mahīmna-stotra:*

asitagirisamam syāt kajjalam sindhupātre surataruvaraśākhā lekhanī patram urvī likhati yadi gṛhītvā śāradā sarvakālam tad api tava guṇānām īśa pāram na yāti.

Freely rendered what the verse means is:

"O Lord Śiva, even if Śāradā, the Goddess of Learning, with the pen of a branch of a divine tree and with an ocean-ful of ink goes on writing all your great qualities on the slate of the Earth for all times to come, she will not be able to reach the other end", (meaning thereby she also will not be able to prepare a complete, exhaustive list of the qualities of Lord Śiva's. Is not the same description of Lord Śiva)greatness applicable to that of the greatness of number-system?

This also reminds us of the Great Puruṣa of the Ṛgveda, who, even after pervading the whole space, remained a little more; *cf.* RV. 10.90.1: *sa bhūmim viśvato vṛtvā' tyatiṣṭhad daśāṅgulam.* The use of the word *daśāṅgulam (=daśa + aṅgulam)* is notable; the author has not used any other number-word there like *pañcāṅgulam,* or *aṣṭāṅgulam* or *śatāṅgulam* etc.

20

The Base 10

We have seen in the fore-going pages that the Vedic texts taken here for study exhibit the knowledge of a full-fledged number-system on which the whole mathematics of the modern age is based. It was a system based on ranks which we indicated by numerals. The system consisted mainly of the original nine numbers from 1 to 9, which we call units. and which the ancient Indian mathematical called the *ekam* numbers (*cf. Līāvati* 1.5); their place was at the right-most position in writing. The basic nine unit steps ran up each successive level of the first rank which was indicated by the figure 1; thus the numbers assumed from first rank onwards the appearance of a 'compound number' as against the non-compound appearance of the basic units. The units thus imparted numbers to the ranks. As in the case of the *dvandva* - compounds in grammar, which could theoretically be expanded upto infinity, the Vedic number-system also theoretically could be expanded to infinity with the help of these ranks. These ranks are indicated by position.

The Vedic system of numerals was purely a decimal system, in which every succeeding rank took a turn at the number *daśa* (10) or its multiples. There is no evidence of any radix other than 10. It does not show any evidence of any other type of number-system based either on the radix of 12 or 20 or 60. The evidences like *aṣṭau* showing 'the dual' (*au* is a suffix for dual in Sanskrit) and referring to 'two fours' etc. are not convincing evidences which go to show 'the radix 4'.[53] Obviously because *aṣṭa* never meant '4'. Such words lead us nowhere. There are references where a number is indicated by a group of other numbers; see for example, 10 = two fives; or 21 = three sevens etc. But we are not very much justified convincingly in stating that in the above cases 5 or 7 are the original bases or radixes and the later numbers like 8, 9, 10 are added later on. Even in the Mayan vigesimal number-system with the base 20, instead of using 19 different unit designations, the system had names only for units from 1 to 10 and used these 10 units only to form the remaining necessary units from 11 - 19 (*cf. K. Menninger, ibid* . p.60). The sexagesimal number-system of the Babylonians also developed as an additional number-system in addition to a decimal number-sequence. The decimal system prevailed in course of time over the sexagesimal system. Also, it is to be specially noted that the sexagesimal system used the decimal number-units from 1-10 only. It had not invented 59 different unit designations. Like the vigesimal system, the sexagesimal system was also interpreted in terms of the units in the decimal system. 10 was acknowledged as the base in Chinese also. Anything incompelete is referred to as "not reaching 10"; and anything in excess is referred to as "12 parts "(*M. Menninger, ibid,* p.84). It is of special interest to note, however, that besides the three number-systems, the Vedic, the Mayan and the Babylonian, which are based on the radixes respectively 10, 20 and 60, there are noted other number-systems of the Indian tribes in North America which exhibit, besides the decimal and vigesimal systems, also the quinary (with the radix 5), ternary (with the radix 3), quaternary (with the radix 4) and octonary (with the radix 8) systems.[54] But it is doubtful, however, whether these tribes have made any progress in building up number-series to higher ranks. The ancient Greek had duodecimal number-system with the base 12. In spite of the existence of different systems, the decimal system of the Vedas has not only outlived all but thrown all the them into total oblivion. The whole modern mathematics is based on the decimal system of the Vedas. The number 10 "plays the part of a threshold or landing;" so all other multiples of 10. "In the Middle Ages the 10-steps and every subsequent multiple of 10, such as 30,80 or 1960, was called *articulus*" (**K. Menninger,** *ibid.* p.45).

20.1. Why the base 10?

Every Vedic number-series takes a turn at the threshold of 10 or its multiples. Why? The general explanation that is advanced is that since man has 10 fingers, 5 on each hand, the number 10 has been found to be the most suitable threshold.[55]

The radix 20 is arrived at by combining the fingers of the hands with the toes of the two feet; thus, 10 finger + 10 toes = 20.

The KS 13.7 and MS 2.3.5 state that there are 10 *prāṇas* (*cf. daśa hi ātman prāṇāh* or simply *daśa prāṇaḥ*). We have also the statement, *nava hi prāṇaḥ; ātmā daśa* (cf. or *nābhir daśami,* KKS.31.13) The *virāj* metre consists of 10 syllables, which are equal to 10 fingers. Hence, since man has 10 fingers, the *virāj* metre is equal to man; *cf.* KS.36.7: *vairājāḥ purṣaḥ*. TS. while explaining the importance of *virāj* says (7.3.9) that 20 are equal to two *virāj* metres; *cf. ete. virājau — viṁśo vai purruṣaḥ* (= 20 are equal to man)[56], because *daśa hastyāḥ aṅgulayaḥ daśa padyāḥ.* There are a number of references which explain 10 and 20 as the sum of the fingers and the toes. The fingers are also called *kṣip* in RV. (3.23.3 etc.) and *svasṛ* (= sisters) which are always in ten.

Besides the fingers, toes and *prāṇa*, the Vedas explain the importance of *daśa* on the ground that a foctus to come out of the mother's womb takes 10 months; *cf.* RV. 5.78.9: *daśa māsān śaśayānaḥ kumāraḥ — nir ā etu;* The adjective *daśamāsya* for the foctus is significant in this respect; *cf* RV. 5.48.7;8 and also other *saṁhitāas*. So we find that besides resorting to finger-counting and body-counting, the Vedas took the help of other natural things or facts for counting. Hence, the *prānas* or *aṅgulayaḥ* or *māsāḥ* or even *yajñāyudhāni* (= the sacrificial instruments which are 10 and are enumerated is TS 1.6.8) - any of these might have served as the principle for taking 10 as the radix. But the truth seems to be that man has come to accept 10 as the base after a great deal of experimentation with different bases like 3, 4, 5, 8, 12, 20 or 60. The Vedas, however, without any hesitation accept the base as 10 and do not even refer to other bases.

To say that *aṣṭau* suggests the base 4 or that *viṁśo* vai puruṣaḥ indicates the base 20 is seeing too much meaning in the statements, because a full-fledged number-system does not seem to have evolved on bases other than 10 neither in the Vedas nor in any other mathematical systems like those of the Babylonian, Egyptian, Mayan or Chinese.

Besides the above references which give us the rationale for the number *daśa* as the base, we have other references from the Vedas which emphasize the importance of the number *nava* as the last limit of the single-digit series from *eka* to *nava*; in an indirect or implied way, the references suggest that the number *daśa* or 10 is the beginning of the new series of two digits. The KKS 31.13, TS 7.5.15 (*nava vai puruṣe trāṇaḥ, nābhirr daśami.*) says: there are nine life-breaths in a man; the navel is the tenth. It shows that the nine *prāṇas* form one category with *nābhi*, the tenth, forming the second. We have also the reference to nine directions. TS.7.1.15 mentions only the nine directions, viz. *prācī* (east), *pratīcī* (west) *dakṣina* (south), *udīcī* (north), the four *avāntara* i.e. middle directions and lastly the *ūrdhvā* referring to the zenith of the sky. The number of directions being ten seems, therefore, to be a post-vedic or rather post-saṁhitā development. These passages can be interpreted to mean that the number *nava* i.e. nine is taken as the last limit of the single-digit series. The number *daśa* by implication, therefore, forms the initial or basic number for the next two-digit series.

21

The Concept of Position

If we study all the possible ways in which the different number-systems are written, we find four main types:- 1. writing by simple grouping system; 2. writing by multiplicative grouping system; 3. writing by ciphered numeral system, and lastly, 4. writing by positional number system.[57]

The system of writing by simple grouping was adopted by the Egyptian hieroglyphics. It accepts 10 as the base. The relation between two types of symbols is of addition.

The multiplicative grouping system differs a little from the above simple grouping system. The relation between the symbols is both of, first multiplication and then addition. This system of writing numerals is found in the traditional Chinese-Japanese system of writing.

The ciphered numeral system of writing was adopted by the so-called Ionic, or alphabetic, Greek numeral system. Other ciphered systems are the Egyptian hieratic and demotic, Coptic, Hindu Brāhmī, Hebrew, Syrian and early Arabic. The last three are alphabetic ciphered numeral systems. It requires many symbols to be memorized in this system. The system of writing numerals on the basis of assigning positions in ascending ranks by ten is the current system of writing numbers, which is borrowed from India. And since Indian system of numbers is based on the Vedas, the Vedic system of numbers also seems to be based on positional values, i.e. values differing according to the position - prior or posterior, left or right — of the numbers. The Babylonian and the Mayan systems are also based on the principle of position or place-value. This system of writing numerals has the advantage of having the least number of symbols, besides being highly convenient.

The Vedic number-system combines in itself two characteristics of the two systems viz. the multiplicative grouping system and the positional numeral system. When, for example, the Veda uses in speech the phrase, *ṣaṣṭim sahasrā navatim nava* (= sixty thousand ninety nine = 60099), the relation between *ṣaṣṭim* and *sahasra* is of multiplication as 60 x 1000; the whole phrase then taken together is added and thus exhibits the relation among its numeral constituents of addition; thus, the whole number 60099 = *ṣaṣṭim* × *sahasrā* (= sixty × thousand) + *navatim* (= ninety) + *nava* (nine). In writing, however, it exhibits the principle of place-value. The numbers to the left show the ten times higher value than the numbers to the right.

The most important difference between the positional numeral system and the other systems is that while the symbols in the latter systems stand for or signify the numbers, the symbols in the former signify, besides the numbers, also the ranks of the numbers in the scale of *daśa*. Thus, the left-hand number 9 indicates its place in the tens while the right-hand number 9 signifies only the single digit 9 in the number-symbol 99.

An important characteristic of the positional numeral system is that it contains or requires the least number of symbols or number-words with the help of which any number can be written. It contains only 20 symbols for numbers and consequently only 20 number words. They are ; *eka, dvi, tri, catur, pañca, ṣaṭ, sapta, aṣṭa, nava, daśa, viṁśati, triṁśat, catvāriṁśat, pañcāśat, ṣaṣṭi, saptati, aśīti, navati, śatam* and *sahasra*. There is, however, no word for zero in the Vedas taken here for study.

But where does this idea of positional notation come from? The Vedas are the oldest literature. As such, we do not have any pre-Vedic evidence which may serve as the source for the Vedic idea of positional notation of numbers. In order to get the answer, we have to take recourse to the linguistic studies in ancient India.

It is generally supposed that linguistic and grammatical activities in ancient India developed later than Veda. But this is only half truth. The whole truth is that Veda is basically a linguistic manifestation of thought. As such, the study of Veda must start from the study of its language. Patañjali is his *paspaśāhnika* bluntly says that of the six[58] vedāṅgas, *Vyākaraṇa* or grammar is the chief auxiliary of the Veda; *cf.* Patañjali: *pradhānam ca ṣaṭsv aṅgeṣu vyākaraṇam.* The origin of the linguistic and grammatical studies, therefore, goes as back as the Vedic times; nay, the linguistic studies must co-incide and be contemporary with the Vedic studies; both of them must go hand is hand. Without linguistic studies, Veda could not have been studied.

What was the form of the Veda in the beginning? Veda or in general language in the beginning, as the Vedic and grammatical tradition in India goes, was *a-vyākṛta*, 'un-analysed' or 'un-described' form. Nobody could understand it. It was a whole linguistic entity. The gods then requested Indra to analyse it and Indra together with Vāyu analysed it; *cf.* TS 6.4.7: *vāg vai parāci avyākṛtā āsīt. te devāḥ indram abruvan, 'imām no vācam vyākuru' iti. ... tām indro madhyataḥ avakramya vyākarot. tasmād iyam vyākṛtā vāg udyate.* Since the time Indra analysed the *vāk* i.e. language referring to the Veda, we, the human beings started speaking an analysed language, analysed into *prakṛti, pratyaya, pada, vākya* etc. The sage Vyāsa, also called Vedavyāsa later on divided the Vedas into different saṁhitās and recensions (*cf.* Mahābhārata, Ādiparva, 17.57 = *vivyāsa*[59] *vedān yasmāc ca tasmād vyāsa* iti smṛtaḥ). The whole Veda then was analysed into, first the *padas* i.e. words, which is known as the *padapāṭha;* the *padas* were analysed further into *prakṛti* (the nominal and verbal base) and *pratyaya* (the suffix and termination); these *prakṛti*-pratyayas were again finally analysed into *varṇas* or sounds. The sounds were thus the last limit and unit of *vāk* in the process of analysis. All the schools of grammar in ancient India has followed this process.

If the above tradition and process of speech-analysis is to be honoured - and there is no reason why it should not be honoured —, the element of position of the spoken sounds and words in terms of time and space must also inevitably be accepted, since obviously all the sounds of a word or all the words in the sentence cannot be spoken all at a time and in the same articulation and places of articulation. They must be spoken one by one or one after the other in order. There is thus an interval of time and space between any two sounds or words. Even Bhartṛhari, who advocates the theory of the simultaneous 'explosion' (*sphoṭa*) of the *śabda* and *artha* and the final unity of both (*cf.* VP. 1.1 and 1.2: *ekam eva yad āmnātam* etc.) on the theoretical plane, had to accept a kind of *krama* i.e order or position of the sound and words on practical plane; *cf.* VP 1.48; *nādasya kramajanyatvād... akramaḥ kramarūpeṇa bhedavān iva jāyate.* When, therefore, the *vāk* which was undivided and un-analysed in the beginning comes to take over an analysed and divided form, the concept of prior and/ or posterior position in time and space does naturally become the chief characteristic of the language-description. And since number-words are words i.e. linguistic entities primarily, they could not be described without considering the prior/posterior positions of *prakṛti, pratyaya, āgamas* and *ādeśas*. This is about simple, non-compounded word-structure.

The problem of position has to be faced prominently especially when one comes to the stage of describing the compound word-structures. Though compounds are composed of at least two members, the constituent members have to be in a fixed order; they cannot be arranged in any order one likes. If the order is changed, the resulting compound gives out a totally different meaning. If, suppose, the order of the two constituent members *rājan* and *puruṣa* in the compound *rāja -puruṣa* (meaning 'the king's servant') is changed to *puruṣa-rāja,* the latter resulting compound structure conveys a totally different meaning from that of the former one; the compound *puruṣarāja* means 'like a king among the males'. Thus the 'position' occupies an important place in considering and describing the compound structures in Sanskrit.

The idea of position, therefore, seems inherent in the spoken words themselves and their description and cannot be taken as totally unknown to the Vedas and ancient Indian linguists. We have in the Veda words which describe the position of an entity or point with reference to another entity; *cf.* words like *puras, purastāt* (= to the east or in front of), *paścāt or paścāttāt* (= to the west or behind), *dakṣiṇa* or *dakṣiṇāttāt* (= to the south or to the right), *uttara* or *uttarāttāt* (= to the north or above), *ūrdhva* or *ūrdhvāttāt* (= above), *adhas* or *adhastāt* (= below), *pūrva* (= before), *para* (= after) and so on. The linguistic studies on Veda like the Nighaṇṭu, Nirukta, the Prātiśākhyas and Pāṇini's Aṣṭādhyāyī - all have very carefully taken into consideration and defined the positions of the sounds, words, *prakṛti, pratyaya, āgama, ādeśa* etc. while describing the Vedic language.

This fact clearly shows that the Veda and linguistics in ancient India were working hand is hand. And the question, who borrowed from whom, is irrelevant and does not arise in view of the facts. If the above question is to be answered at all, the only answer seems to be that lingustics or grammar in ancient India borrowed from Veda the concept of position or place-value, since Vedas are earlier than all the extant grammars of Sanskrit. The only difference between the Vedic and linguistic understanding of the concept of position seems to be that the linguisticians or grammarians made explicit reference to and explicity defined the concept at the time of analysing and describing the Vedic language whereas the concept was latent or implicit in the Veda.

Although the Veda, as we have seen above, exhibits two types of languages- mathematical and non-mathematical - depending on whether a word conveys a numerical meaning or not, this distinction is absolutely immaterial for a linguist because for him the word is a word to be described irrespective of the fact whether it conveys a mathematical or non-mathematical meaning. Hence, as the concept of position was regarded as very important in the case of the description of non-mathematical language/word, so also it was equally important in describing the mathematical language/word. And there seems to be nothing wrong in such a methodology. Hence the concept of position was employed in describing the mathematical language/word also. Just as, therefore, in a non-mathematical compound like *vajra-bhṛt,* the word *vajra* occupied the prior place and the word *bhṛt,* the posterior one, similarly in the mathematical compound like *ekā-daśa* the word *eka* occupied the prior position and the word *daśa*, a posterior one.

Upto this i.e. descriptive stage it was all right. But the problem arose when they came to the stage of assigning symbols to the number-words spoken for purposes of writing. The problem was of the type of symbols to be assigned— whether or not to assign the same symbols for both the types of languages - mathematical and non-mathematical. Obviously, since the two languages signified two different types of semantic levels/categories and two different worlds of thought they opted for assigning different symbols for the two different types of languages.

But then, there was another problem of the position or order of the symbols also—whether to follow the same order which was available in the spoken language for both the types of languages or to make a distinction between the two? At this stage, the Vedic people opted for the second alternative and the order of the symbols for the non-mathematical language/word preserved the original order of the spoken words while the order of the mathematical symbols was reversed. Thus, in the non-mathematical compound *vajrabhṛt*, the symbols for the two components viz. *vajra* and *bhṛt* retained their original order in which they were spoken (i.e. *vajra* in the first place and *bhṛt* in the second place) whereas in the mathematical word-compound viz. *ekādaśa*, the symbols for the two components (i.e. *eka* and *daśa*) reversed their positions with *daśa* in the first place and *eka* in the second place. This is what is meant by *aṅkānām vāmato gatih*, 'the order is to be reversed in the case of numbers'. Thus, though this rule for the order of the numerical symbols might have been explicity spelt later on i.e. later than the Veda, it certainly seems to have been very well-known in the times of the Vedas. Also, the rule does not indicate the beginning of writing, but pre-supposes writing the mathematical and non-mathematical languages in terms of symbols. And since the number-words were transformed into number-symbols by writing in the reverse way in the times of the Vedas themselves (the way which we are still following), the rule simply stated the principle and fact of transforming the number-words into number-symbols of figures. The rule thus is a descriptive rule and not a prescriptive one; but it came later on to be accepted as a prescriptive one, which is still followed upto this day not only by Sanskrit writers but by all the civilisations of the world, old and modern. And we have the way of writing the numbers as (to exemplify the first series):

11, 12, 13, 14, 15, 16, 17, 18, and 19.

The number-symbol to the left stand for the serial number of the series and those to the right for the unit-numbers. The prior number-symbol which is common to all indicates the first number of the series.

If, suppose, the rule *aṅkānām vāmato gatiḥ* would not have been there and the Vedic people would have opted to follow the same order of the number word-compounds in the case of the number-symbols also, the first series would have looked like the following:

11, 21, 31, 41, 51, 61, 71, 81, and 91,

in which case, the number to the right would have denoted the serial number of the series, the units occupying the first position. In that case, the whole mathematical system of today would have basically been totally different not only in appearance but also in calculation and computing from the present one. And 11+1 would have given us not 12 but 21 and 21+1 would have been equal to not 22 but 31, had not there been the Vedic system of number-writing and the implied rule *aṅkānām vāmato gatiḥ* ; ten would have been not 10 but 01 and hundred, not 100 but 001.

We can thus see that the concept of position seems to have been accepted and resorted to right from the times of the Vedas and the rule *aṅkānām vāmato gatiḥ* leads us to assume the existence of writing in Vedic times.

22

The Journey of Zero

The *Amarakoṣa* notes the following words for zero; *asāram phalgu śūnye ca vaśikam tuccha-riktake,* (*Amarakoṣa* 3.56). Later on in post-vedic times, the numbers came to be designated by symbolic words and the number zero came to be signified by words like *śūnya, kha, ākāśa, ambara, vyoma, nabhas, pūrṇa* etc. majority of which are the words for *ākāśa*, 'sky'. The Vedic texts, however, have not used any number-word for zero. The first occurrence of the word *śūnya* is found in AV.14.2.19 (*śūnyaiṣī nirṛte yājagandhā uttiṣṭha*) in which the goddess Nirṛti is addressed as *śūnyaiṣi. = śūnya + eṣī,* 'desiring or looking for vacant places or houses'. The word *śūnya,* therefore, means 'nothing, void, vacant or un-occupied place' etc.

The word *śūnya* is derived by Kṣīrasvamī, the commentator of Amarakoṣa as follows:-

śune hitam śunyam śūnyam ca. ugavādibhyo yat ity atra samprasāraṇam vā ca dīrghatvam.

There is no mention of the verbal root here from which it is derived. Yet, the verbal roots seems obviously the root *śvi* 'to go, to increase' etc. (*cf.* Pāṇinian *dhātupatha, ṭu-o-śvi gativṛddhyoḥ).* Thus, formally speaking, the root *śvi* with the past passive participial suffix *ta* > *na* gives out the form *Śuna*, meaning 'that which has increased or gone'. To *Śuna*, the suffix *-ya* is applied according to the Pāṇinian sūtra, ugavādibhyo yat, 5.1.2 The *gaṇapāṭha* does not include the word *Śuna* in ugavādigaṇa; yet the *gaṇasūtra* derives it by passing the remark quoted above by Kṣīrasvāmī. Bhaṭṭoji Dīkṣita derives two words, with and without long *ū* as *Śunya* and *Śūnya*.

Semantically, however, the words *Śunya* or *Śūnya* donot convey the sense of 'vacant, void, un-occupied etc, though they became current in later Sanskrit indicating zero. Derivationally, the word shares its etymology with the word *Śvan* or *Śunaka* meaning 'dog'. And since the dog is/was a despised animal, not worth-attention, the derived word *Śunya* or *Śūnya* seems to have come to signify the sense of 'something useless, worth-throwing away' and later for 'zero in mathematics indicating 'a neglible number'; *cf.* the word *tuccha* 'rubbish' noted by Amarakoṣa.

We have seen before that the idea of positional notation and consequently as its corollary the concept of zero and its positional value are available in the Vedic texts proper. They are not the post-Vedic ideas. And the ease and frequency of the reference to these concepts in the Vedic texts take us far back again to pre-Vedic contributions to mathematics of the world. It is from the Vedas that the ideas spread outside India and abroad. The Vedas are the lenders and not borrowers.

As it sometimes happens, the Sanskrit word *Śūnya* is not borrowed phonetically by the outside civilisations and people like the Arabs, who were the first borrowers of the concept and word for zero and the numerals. Though the numerals are called the Arab numerals in the West, they are not invented by the Arabs proper. They are Vedic inventions. What the Arabs did was that they translated in their language the idea of 'absence, void, un-occupied place, nothingness' which was conveyed by the Sanskrit word *Śūnya*. The Arab word for vacant, empty was '*as-sifr*' or more accurately *aṣ-ṣifr* (the *ṣ* here is a palatal as in Sanskrit *Ś* in *Śūnya*).

From the Arabs the word got its way to the West in all its two aspects, viz.. the symbol as well as its name. The *ṣifra* of the Arabic changed to *cifra* in Latin and to *cephirum* in Greek. The Latin *cifra* was transformed phonetically into *zefiro, zefro* or even *zevero* in Italy, which was shortened by the loss of the middle sounds *f* and *ve* to the form zero. The French has *chiffre* and the German, *Zeiffer and cifra.* English has *zero.*

The concept of place-value notation and the zero is a gift of the Vedas to human mathematics. The zero especially was taken in the west to be 'a creation of the Devil', since it itself is nothing but increases ten times the value of the number against which it is put or read. The following quotation from **Karl Menninger's** book (*bid.* pp. 422, 423, 424) will give an idea about the embarrassment caused by zero in the beginning of the Middle Ages in the West:

"Computations with the new numerals, in contrast, were certainly not as easy to visualize. But most important of all they embodied an intellectual obstacle that was scarcely overcome during the first few centuries of their presence in the West : the zero!

The Zero Again

What kind of crazy symbol is this, which means nothing at all. Is it a digit, or isn't it? 1, 2, 3, 4, 5, 6, 7, 8 and 9 all stand for numbers one can understand and grasp - but 0? If it is nothing, then it should be nothing. But sometimes it is nothing, and then at other times it is something; 3 + 0 = 3 and 3 - 0 = 3, so here the zero is nothing, it is not expressed, and when it is placed in front of a number it does not change it: 03=3 so the zero is still nothing, *nulla figura!* But write the zero *after* a number, and it suddenly multiplies the number by ten : 30 = 3 x 10. So now it is somthing - something in-comprehensible but powerful, if a few "nothing" can raise a small number to an immeasurably vast magnitude. Who could understand such a thing? And the old and simple one-place number 3000 (on the counting board) has now become a four-place number with its long tail of "nothing" - in short, the zero is nothing but "a sign which creates confusion and difficulties," as a French writer of the 15th century put it - *une chiffre donnant umbre et encombre.*

Thus the resistance to the Indian numerals by those who used the counting for calculations took two forms: some regard them as the creation of the Devil, while others made fun and ridiculed them:

Just as the rag doll wanted to be an eagle, the donkey a lion, and the monkey a queen, the *cifra* put on airs and pretended to be a digit, wrote an educated man in France as late as the 15th century. According to another French source, an "algorism-cipher" is a term of abuse of the same class as blockhead. Astrologers, however, gladly adopted the new numerals; like every form of secret writing, they helped to raise their status. The Algorism of the Salem Monastery correctly interpreted the new numerals and used them for computations, but they still created such confusion in the mind of their author that he appended the following mystical interpretation:

Every number arises from One, and this in turn from the Zero. In this lies a great and sacred mystery - *in hoc magnum latet sacramentum -:* HE is symbolized by that which has neither beginning nor end; and just as the zero neither increases nor diminishes/another number to which it is added or from which it is subtracted/so does HE neither wax nor wane. And as the zero multiplies by ten/the number behind which it is placed/so does HE increase not tenfold, but a thousandfold - nay, to speak more correctly, HE creates all out of nothing, preserves and rules it - *omnia ex nichillo creat, conservat atque gubernat.'*

In this way the zero acquired its profound "significance" and began to represent something.

But the learned men too were not sure whether the zero was a symbol, a numeral, or not. According to the name *Null* which they gave to it, it was not; and so medieval writers would frequently present the "9 digits" to which they would add one more, which was called a *cifra*: . . . .

He who wishes to learn to reckon with digits must begin by knowing the figures of the digit/and then learn the force and meaning of the place-values according to which the digits are set. And there are nine figures that have value meaning/and one more figure outside of them which is called null, O, which has no value in itself/but increases the value of the others.

Another manifestation of the same confusion and insecurity was the many names given to the zero (see p.401). What was the point of forsaking the old reliable counting board for something so full of contradiction that only a few learned men could understand it, and even they just barely? Even today the expression *faire par algorisme*, "to do it with the algorism," is still used in France in the sense of "to do it with the algorism, "to miscalculate."

This popular disinclination to use the new numerals was also behind the attempt to make these strange new concepts, the zero and the place-value principle, comprehensible by presenting them in verse form; thus Alexander de Villa Dei said of the zero (see p.412) that *cifra nil significát, dat significare* sequentti

the zero has no value, but gives value to the next [digit of higher rank];

and he explained place value in the following lines (of which only the beginning and end are quoted here):

unum dat prima, secunda decem, dat tertia centenum

quarta dabit mille, milia quinta decem ....

chifra nil condit, sed dat signare sequentem,

The first [place] makes [the digits there worth] units, the second tens, the third hundreds, the fourth thousands, the fifth ten thousand .. the zero itself makes nothing, but it makes the following digits have [greater] value."

K. Menninger (*ibid* 401) gives the following figure to show the steps through which the phonetic changes the Sanskrit word *Śūnya* underwent in its journey from East to West through Arabic people:-

Sanskrit (6th to 8th cent.)	śūnya (= "empty"),		
Arabic (9th cent.)	*as-sifr*		
Latin (13th cent.)	*cifra*	*Zefirum*	
French (14th cent.)	*chiffre*	*zefiro-zevero-zero*	Italian.
German (15th cent.)	*Ziffer*	*zero*	French, English.

Thus we find that the concept, word and symbol for zero travelled from India, first to the Arabic countries and then to Italy, South Germany and then to Western European countries France and England. The idea of zero, which is based on place-value notation, is certainly Vedic from which it spread to different countries in the West.[60] The Vedas themselves in their present form and contents presuppose a very, very long history. They represent a mature stage of a civilisation whose roots go back to hoary antiquity.

As the tradition says, the sage Vedavyāsa collected together all the scattered Vedic material and divided it into four *saṁhitās*; *cf. Mahābhārata* (Vol. I, Ādiparvan, 1.57.70, BORI, Pune, 1971):

brahmaṇo brāhmaṇānām ca

tathā' nugrahakāmyayā;

vivyāsa vedān yasmāc ca

tasmād vyāsa iti smṛtaḥ.

This means that the number-system, the device of the place-value notation and the concept of zero, as available in the Vedas, go back to still ancient times of which there is no record. The remarks such as "I can only compare their (= Indians'; brackets mine) mathematical and astronomical literature ... to a mixture of pearl shells and sour dates, or of pearls and dung, or of costly crystals and common pebbles. Both kinds of things are equal in their eyes, since they cannot raise themselves to the methods of a strictly scientific deduction", by Alberūnī (c. 1000), the Arab historian, quoted by **D.E. Smith**[61] are, therefore, totally unwarranted in view of the facts presented here.

23

Resumé

To sum up the whole discussion given in the foregone pages. We are now, it is hoped, in a somewhat better position than before to gather at least a broad idea about the stage of mathematical development in the times of the nine Vedic *samhitās*. The *samhitā* literature taken here for study gives us the following picture about the mathematical knowledge in those ancient times.

23.1. The vedic system of numbers is a full-fledged system based on the positional ranks. If the position of the number is changed, its rank also changes. That the Vedic people knew the concept of position is seen from that fact that the idea of position is inherent in the linguistic analysis itself. Thus, in *pañca-viṁśati*, since the word *pañca* is spoken first, its position is prior to or earlier than the word *viṃśati*.

23.2. The whole number-system pre-supposes linguistic analysis. It is with the help of linguistic analysis of the Vedic language that we get a definite linguistic procedure for arriving at the two-digit notation of *daśa* onwards (upto 99) and hence the zero in *daśa*. The zero in *daśa*, therefore, is the absence of a positive entity in rank no. One from right. Consequently zero in the place of any position indicates the absence of that positional rank. This zero substituted for the absence of the rank has got to be distinguished from the zero as a number below 1, which is obtained by the simple process of substraction of two equal numerical entities such as 5-5 = 0 or in general $x-x = 0$

23.3. Although we do not find any special word for zero, which is the greatest invention of all times of the Indian Vedas, we can easily arrive at the zero with the help of a suitable linguistic procedure. Though while adopting the linguistic procedure we seem to borrow it from the post-Vedić Pāṇinian technique of linguistic analysis and description, the procedure certainly seems to have been known to the Vedic linguists and mathematicians. And the truth seems to be that Pāṇini, who is post-Vedic, seems to have borrowed the technique of linguistic analysis from the Vedic. And once Pāṇini entered the stage of grammar-composition the pre-Pāṇinian Vedic technique seems to have been credited to Pāṇini's name. The Vedas also know full well the maxim *aṅkānām* vāmato gatiḥ, without which the zero in *daśa* symbolised as 10 cannot be explained.

23.4. The mathematical zero can be compared only with the *Pāṇinian* zero, both in the procedure of arriving at zero as well as in its nature as indicating a substitute for something which should have been there, but which is not there. The only difference between the mathematical and Pāṇinian zero is that while the Pāṇinian zero is a substitute for some phonemes and morphemes, and not their place the mathematical zero is a substitute for the place or rank of the number, and not the number itself. Another point of distinction between the two types is that while in actual spoken language or in writing the zero of phonemes or morphemes is not traceable (in the sense that its existence is not felt) the mathematical zero requires to be indicated both in speech and writing. Thus in the form *atti,* the zero of the *vikaraṇa śap* is neither spoken nor written; whereas in the number, say, 105, the zero is conspicuously mentioned or felt both in speech (because while speaking we do not speak the decimal place and speak only 'one hundred and five', indirectly indicating thereby the zero i.e. the empty space in between the two numbers 'one' and 'five') as well as in writing (because we specifically put the symbol for zero, viz. O, between the two numbers). *cf.* for example the Vedic phrases like *śatam ekam ca* (for 101; RV. 1.117.18), *śatam sapta ca* (for 107; RV. 10.97.1), *ṣaṣṭim sahasrā navatim nava* (for 60099; RV. 1.53.9) etc. in which the missing ranks of *daśa* (in the number 101 and 107) and *śatam* (in the number 60094) are hinted by their absence in speech and require to be represented by the symbol for zero viz. O in writing. Of course, one important point cannot be forgotten in comparing both the mathematical the Pāṇinian zero, viz. in both there is an inherent comparison of the mathematical and linguistic expressions with missiong ranks and sounds with those which exhibit their presence. To explain, in the linguistic expression *atti* (without *śap*), there is its inherent comparison with *bhavati* (with *śap*); so also, in the mathematical expression *śatam ekam ca* (without-*daśa* rank) there is its inherent comparison with, say, 125 or 156 or 189 etc. which contain the *daśa* rank. Actually in saying that the places of *daśa* etc. in 101 etc. are vacant, we have in our mind the numbers like 125 etc. in which the ranks of *daśa* etc. are positively filled up by some positive numbers; in numbers like 101, 107 etc. we fill the vacant places with zero—that is the only difference between the two types of numbers.

23.5. The number-series is based on the radix of *daśa* i.e. ten. It is thus a decimal system of numbers. Except some references like *viṁśo vai puruṣaḥ* (= 'man is twenty') in some texts, we do not get convincing evidence of any radix other than ten, like five, six, twelve, twenty or sixty. Either they knew these radices and found them to be not much suitable and convenient and hence rejected them, or they did not have the idea of the existence of any other radices. In the face of the passages like *viṁśo vai puruṣaḥ, saptandaśaḥ prajāpatiḥ* etc., the former conclusion seems to be more rational. This means that the Vedic people must have arrived at the base *daśa* after a great deal of experiment with different radices. The decimal number-system, therefore, seems to have come to be established after a great deal of trial-and-error method.

That the decimal number-system came to be taken as the most perfect for measuring the present structure of the universe is supported by the fact that the same Vedic decimal system has been adopted through all these ages upto the present day. With only 20 numbers or number-words, viz. 1, 2, 3, 4, 5, 6, 7, 8, 9, 10, 20, 30, 40, 50, 60, 70, 80, 90, 100 and 1000, the Vedic people could erect a gigantic number-structure with infinite possibilities of building up infinite number of series. All these 20 wórds are non-compound, underived, word-structures, as against all other infinite number words which are all derived. It is, however, to be noted that though the number word *daśa* is linguistically/grammatically underivable, the symbol for *daśa* viz. 10 as consisting of two digits is mathematically derivable, as we have shown before.

23.6. All the numbers recorded in the nine Vedic *saṁhitās* are positive numbers. There is absolutely no mention of or reference to any negative numbers.

23.7. The Vedic people knew very well the four basic or primary mathematical operations of addition, substraction, multiplication and division.

Instead of using any signs, which practice might or might not have been there in those times, they have used sign-words like *ca, sākam* or sign-suffixes like *s* or *kṛtvasuc* etc. for indicating the different operations. The knowledge of fractions follows as a natural corollary of the process of division. Working out the mathematical operations with high numbers, however, is not seen anywhere in any of the *saṁhitās*.

23.8. The Vedas have given many series of numbers which in modern terminology can be termed as the series of arithmetic and geometric progression. We cannot make out any purpose behind explicity stating in great length all the different series. Can it be that the passages giving the different series are excerpts from some ancient, pre-Vedic texts on mathematics?

23.9. The passages VS 17.2, TS. 7.2 12-19 etc. giving out the different ranks of *daśa* show that the Vedic people knew very well the structuring of the numerical series with the help of what we today call as the exponents or indices. Thus we get, *śata* = *daśa* x *daśa* = *daśa*^{2} ($100 = 10 \times 10 = 10^2$), *sahasra* = *daśa* x *daśa* x *daśa* = *daśa*^{3} ($1000 = 10 \times 10 \times 10 = 10^3$) and so on. Algebraically, if $ś$ = *śatam*, s = *sahasram* and d = *daśa*, we get $ś = d^2$ and $s = d^3$ etc. The Vedic statements, however, require to be studied further algebraically.

23.10. The passages like the VS 18.24, 25 (repeated in other *saṁhitās*) show the knowledge on the part of Vedic people of squares and of the procedure to find out the squares of numbers. We, however, donot get any evidence of the procedure of finding out square-roots.

The passages like *pañcāśate savāhā śatāya svāhā* etc. (TS. 7.2.19 etc.) show that the Vedas know the existence of surds and the geometrical procedure to find out the numerical value of some surds or numbers whose square-root cannot be drawn in terms of full integers, like $\sqrt{50}$, $\sqrt{200}$, $\sqrt{300}$ etc.

The passages like VS 17.2 etc. and TS 7.2.19 etc. contain the seeds of the later algebraic and geometrical considerations. The subject, however, requires a deeper study.

23.11. The phrases like *yajñena kalpantām* or *sarvasmai svāhā* are used to suggest that the same procedure as is adopted in the previous cases should be followed further *ad infinitum* to arrive at the desired results in the series. The fact that the series they have stated in the passages can be expanded to infinity by means of the latent principles shows that they had a clear idea about infinity, as they had about zero. Not only this, but the concept of zero and infinite expansion could be grasped mathematically by adopting a definite mathematical process.

23.12. We do not get any enunciation of any philosophy of mathematics in general and numbers in particular in the Vedas. If possible we are likely to find it in the later literature of the Brāhmaṇas and Upaniṣads. Some grammarions like Kauṇḍabhaṭṭa (*cf. Vaiyākaraṇabhuṣaṇasāra*) and all Naiyyāyikas from Praśastapā da (*cf.* Praśasta*pādabhāṣya*) to *Gadādhara* (*cf. Tattvacintāmaṇi*) seem, however, to have given some thought to the philosophy of *Saṁkhyā* i.e. numbers. But the subject of mathematical philosophy requires to be studied afresh, basing our judgements on sound Vedic evidences.

23.13. As regards writing, we do not get any direct explicit and/ or convincing evidence to conclude that the Vedic people knew the art of writing. Yet if the possible interpretations of the Vedic mathematical data proposed here are any evidence, we may safely conclude that the Vedic people knew the art of writing at least the number-symbols. We meet with the mention of very big numbers consisting of four, five or even ten digits and hence places. Also, we have the process to arrive at zero and infinity. Moreover, the procedure to arrive at the squares of different numbers is also given. The passages like VS 17.2 and Ts 7.2 12-19 exhibit clear knowledge of surds like $\sqrt{50}$, $\sqrt{200}$, $\sqrt{300}$ etc. All these facts lead us to the inevitable conclusion that the Vedic people knew the art of writing. The computation and calculations involved in all the above processes cannot be done without writing the numbers; simply oral calculation cannot lead one to comprehend the higher, abstract, ideas of squares, zero, infinity etc. We donot know what symbols were used for numbers and/or number-words. But that such higher, subtle calculations as are found out to be in the Vedas are not possible and cannot be orally carried out without the help of some written symbols, crude or refined, seems to be the most logical conclusion. The study must be pursued further on the basis of the comparison of the Vedic civilisation with other civilisations like Indus-Valley civilisation, the Egyptioan civilisation, the Babylonian civilisation, the Chinese civilisation and even with the far off civilisation of the Mayas.

Appendix—A

This Appendix A gives the number-words from as many as 20 Indo-European languages. Barring the reconstructed Indo-European, which gives only 12 words, all other 19 languages supply us all the relevant data in full. Of these 19, Sanskrit, Old Greek and Latin are ancient IE languages. French, Italian, Portuguese and Spanish are from the Romance family of languages. The Anglo-Saxon group of languages is covered by the two main languages, viz. English and German. The Russian language is given as a specimen of the Slavic group of languages. The Indo-Iranian picture of the number-words is presented by the Old Persian i.e. Avestan and the New i.e. modern Persian languages. The Prākrita, Pāli, Hindi, Marathi, Gujrati, Oḍia and Bengali give us the picture of the number-words as available in the Middle Indo-Aryan and New Indo-Aryan stage. All these words have followed a very long course of history of about at least five thousand years. As such, they require to be studied from the point of view of Historical Linguistics, although much work has been done in the direction by philologists. The aim of this Appendix A is to present a consolidated picture of the number-words at a glance, to facilitate their further study from synchronic and diachronic point of view.

1. Ancient Languages

Modern Symbols	*Indo-European Number-Words*	*Old Greek*	*Latin*	*Sanskrit*
1.	**Oi*	*eis*	*unus*	*eka*
2.	**duo*	*duo*	*duo*	*dvi*
3.	**tri*	*treis*	*trēs*	*tri*
4.	**qwetur*	*tetra*	*quattuor*	*catur*
5.	**penqwe*	*pente*	*quinque*	*pañca*
6.	**sweks*	*esh*	*sex*	*ṣaṣ*
7.	**septm*	*hepta*	*septem*	*sapta*
8.	**oktō*	*okto*	*actō*	*aṣṭa*
9.	**newn*	*ennea*	*novem*	*nava*
10.	**dekṃ*	*deka*	*decem*	*daśa*
11.		*endeka*	*undecim*	*ekādaśa*
12.		*duodeka*	*duodecim*	*dvādaśa*
13		*treisdeka*	*tredecim*	*trayodaśa*
14.		*tetradeka*	*quattuordecim*	*caturdaśa*
15.		*pentedeka*	*quindecim*	*pañcadaśa*
16.		*eshdeka*	*sēdecim*	*ṣoḍaśa*
17.		*heptadeka*	*septendecim*	*saptadaśa*
18.		*oktodeka*	*duo-dē-vīginti*	*aṣṭādaśa*
19.		*enneadeka*	*undēvīginti*	*navadaśa/ ekonaviṁśati*
20.	**Kṃt*	*koti*	*vīginti*	*viṁśati*
21.		*eikoti*	*vīgintī-unus*	*ekaviṁśati*
30.		*triakonta*	*triginta*	*triṁśat*
40.		*tetrakonta*	*quadraginta*	*catvāriṁśat*
50.		*pentakonta*	*quinquaginta*	*pañcāśat*

60.		*hexakonta*	*sexaginta*	*ṣaṣṭi*
70.		*heptakonta*	*septuaginta*	*saptati*
80.		*oktukonta*	*octoginta*	*aśīti*
90.		*enneakonta*	*nonaginta*	*navati*
100.	**Kṃ̥tO*	*hekaton*	*centum*	*śatam*
1000.		*xilioi*	*mille*	*sahasra*

2. The Romance-Group of Languages

Modern Number-Symbols	*French*	*Italian*	*Portuguese*	*Spanish*
1.	*un*	*uno(m), una(f).*	*um, uma*	*uno*
2.	*deux*	*due*	*dois, duas*	*dos*
3.	*trois*	*tre*	*três*	*tres*
4.	*quatre*	*quattro*	*quatro*	*cuatro*
5.	*cinq*	*cinque*	*cinco*	*cinco*
6.	*six*	*sei*	*seis*	*seig*
7.	*sept*	*sette*	*sete*	*siete*
8.	*huit*	*otto*	*oito*	*ocho*
9.	*neuf*	*nove*	*nove*	*nueve*
10.	*dix*	*dici*	*dez*	*diez*
11.	*onze*	*undici*	*onze*	*once*
12.	*douze*	*dodici*	*doze*	*doce*
13.	*treize*	*tredici*	*treze*	*trece*
14.	*quatorze*	*quattordici*	*catorze*	*catorce*
15.	*quinze*	*quindici*	*quinze*	*quince*
16.	*seize*	*sedici*	*dezasseis*	*deiciseis*
17.	*dix-sept*	*dicias-sette*	*dezassette*	*diecisiete*
18.	*dix-huit*	*dici-otto*	*dezói+ito*	*dieciocho*

19.	*dix-neuf*	*dici-annove*	*dezanove*	*diecinueve*
20.	*vingt*	*venti*	*vinte*	*veinte*
21.	*vingt-et-un*	*ventuno*	*vinte e um*	*veintiuno*
30.	*trente*	*trenta*	*trisenta*	*treinta*
40.	*quarante*	*quaranta*	*quarenta*	*cuarenta*
50.	*cinquante*	*cinquanta*	*cincoenta*	*cincuenta*
60.	*soixante*	*sessanta*	*sessenta*	*sesenta*
70.	*soixante-dix*	*settanta*	*setenta*	*setenta*
80.	*quatre-vingts*	*ottanta*	*oitenta*	*ochenta*
90.	*quatre-vingt-dix*	*novanta*	*noventa*	*noventa*
100.	*cent*	*cento*	*cem*	*ciento*
1000.	*mille*	*mille*	mil	*mil*

3. The Anglo-Saxon Group The Slavic Group

Modern Number-Symbols	*German*	*'English*	*Russian*
1.	*ein, eins*	*one*	*adin*
2.	*zwei*	*two*	*dva*
3.	*drei*	*three*	*tri*
4.	*vier*	*four*	*cetire*
5.	*fünf*	*five*	*pyat*
6.	*sechs*	*six*	*shest*
7.	*sieben*	*seven*	*sem*
8.	*acht*	*eight*	*vosem*
9.	*neun*	*nine*	*devyat*
10.	*zehn*	*ten*	*desyat*
11.	*elf*	*elevan*	*adinnadtsat*
12.	*zwölf*	*twelve*	*dve-adtsat*
13.	*dreizehn*	*thirteen*	*tri-adtsat*

14.	*vierzehn*	*fourteen*	*cetir-adtsat*
15.	*funfzehn*	*fifteen*	*pyat-adtsat*
16.	*sechzehn*	*sixteen*	*shest-adtsat*
17.	*siebzehn*	*seventeen*	*sem-adtsat*
18.	*achtzehn*	*eighteen*	*vosem-adtsat*
19.	*neunzehn*	*nineteen*	*devyatnadtsat*
20.	*zwanzig*	*twenty*	*dvadtsat*
21.	*einundzwanzig*	*twenty-one*	*dvadtsat-odin*
30.	*dressig*	*thirty*	*tridtsat*
40.	*viersig*	*forty*	*sorak*
50.	*funfsig*	*fifty*	*pyatdesyat*
60.	*sechzig*	*sixty*	*shestdesyat*
70.	*siebzig*	*seventy*	*semdesyat*
80.	*achtzig*	*eighty*	*vocemdesyat*
90.	*neunzig*	*ninety*	*devyanasto*
100.	*hundert*	*hundred*	*sto*
1000.	*tausend*	*thousand*	*tisyaca*

4. The Indo-Iranian Group of Languages.

Modern Number-Symbols	*Avesta/Old Persian*	*New Persian*
1.	*ai-va*	*yak*
2.	*dva/duva*	*do*
3.	*øri/ørai*	*se*
4.	*catur*	*cahar*
5.	*pañca*	*pañj*
6.	*xśvaś*	*sed*
7.	*hapta*	*haft*
8.	*aśta*	*haśt*

9.	*nava*	*noh*
10.	*dasa*	*dah*
11.	*aiva-dasa*	*yazdah*
12.	*duva-dasa*	*davazdah*
13.	*ørai-dasa*	*sizdah*
14.	*caøru-dasa*	*cahardah*
15.	*pañca-dasa*	*panzdah*
16.	*xšavas-dasa*	*sanzdah*
17.	*hapta-dasa*	*hevdah*
18.	*ašta-dasa*	*hizdah*
19.	*nava-dasa*	*nuzdah*
20.	*vīsaiti*	*bist*
21.	*aiva-vīsaiti*	*bistoyak*
30.	*øri-sat*	*si*
40.	*caøvar-sat*	*cehel*
50.	*pañca-sat*	*panjah*
60.	*xšvašti*	*shast*
70.	*haptāti*	*haftad*
80.	*aštāti*	*hashtad*
90.	*navati*	*navad*
100.	*sata*	*sad*
1000.	*hazañghra*	*hazār*

5. The Indo-Aryan Group of Languages

Modern number-Symbols	*Prākrita*	*Hindi*
1.	*eka, ega, ekka, ego, eo*	*ek*
2.	*du, donni, do, due, be, duye*	*do*
3.	*tiṇiṇa, tinni*	*tīn*
4.	*cattāri, cattaro, cauro*	*cār*

5.	*pāñca, pāca.*	*pā̃c*
6.	*cha*	*chah*
7.	*satta*	*sāt*
8.	*aṭṭha*	*āṭh*
9.	*nava, ṇava, na-a, ṇa-a*	*na-v, nau*
10.	*dasa, daha, ḍaha, raha*	*das*
11.	*eāraha, gyāraha*	*gyāraha*
12.	*bāraha, bārasa*	*bāraha*
13.	*teraha, terasa*	*teraha*
14.	*ca-ud-daha, cauddasa*	*caudaha*
15.	*paṇṇarasa, paṇareha, paṇaraho, paṇāraho*	*pandraha*
16.	*solaha, solasa*	*solaha*
17.	*sattaraha, sattarasa*	*sattarah*
18.	*aṭṭhāraha, aṭṭhārasa*	*aṭhāraha*
19.	*unvīsai, unavīsā, eknavīsā*	*unnīs*
20.	*visata, vīsai, vīsā*	*bīs*
21.	*ekavīsā*	*ikkis*
30.	*tīsā, tisa-ā, tīse*	*tīs*
40.	*cattālīsā*	*cālīs*
50.	*pañcāsā, paṇāsā, pannā*	*pacās*
60.	*saṭṭhī, saṭhṭhī*	*sāṭh*
70.	*sattari, sayari, sattaras*	*sattar*
80.	*asī-ī*	*assi*
90.	*navvae*	*nabbe*
100.	*sata, saya, sa-a, sa-ā*	*sau*
1000.	*sahassa*	*hazār*

	Marathi	*Gujrati*	*Odia*	*Bengali*
1.	ek	ek	eka	ek/aek
2.	don	be	dui	dui
3.	tin	traṇ	tini	tin
4.	cār	cār	cāri	cār
5.	pāc	pā̃c	pāñca	pāc
6.	sahā	cha	cha	choy
7.	sāt	sāt	sāta	sāt
8.	āṭh	āṭh	āṭha	āṭh
9.	na-u	nav/nau	na	noy
10.	dahā	das	daśa	doś
11.	akrā	agyār	egāra	egāro
12.	bārā	bār	bāra	bāro
13.	terā	ter	tera	taero
14.	caudā	caud	ca-u-da	coddo
15.	pandhrā	pandhar	pandara	ponero
16.	soḷā	soḷ	ṣohala	śolo
17.	satrā	sattar	satara	śotero
18.	aṭhrā	aḍhār	aṭhara	āṭhāro
19.	ekoṇīs	uguṇīs	uṇeisi/uṇeiśa	uniś
20.	vīs	vīs	koḍie	bīs/kuri
21.	ekvīs	ekvīs	ekoisi	ekus
30.	tīs	tīs	tiriśa	tiriś
40.	cāḷīs	cālīs	cāḷiśa	collīś
50.	pannās	pacās	pacāśa	poncās
60.	sāṭh	sāṭh	ṣāṭhie	śāṭh
70.	sattar	sittyer	saturi	śotter

80.	*āiṁśī*	*assī*	*aśī*	*āśī*
90.	*nawad*	*ne-ū*	*nabbe*	*nobboi*
100.	*śambhar*	*so*	*śaha*	*śo/śoto*
1000.	*hajār*	*hajār*	*hajāra*	*hājār*

Pāli

1. *eka*
2. *dvi*
3. *ti*
4. *catu*
5. *pañca*
6. *cha*
7. *satta*
8. *aṭṭha*
9. *nava*
10. *dasa*
11. *ekādasa/ekārasa*
12. *dvadasa/dvārasa/bārasa*
13. *tedasa/terasa/teḷasa*
14. *catuddasa/cuddasa/coddasa*
15. *pañcadasa/paṇṇarasa*
16. *soḷasa*
17. *sattadasa/sattarasa*
18. *aṭṭhādasa/aṭṭhārasa*
19. *ekūnavīsati/ekūnavīsā*
20. *vīsati/vīsā*
21. *ekavīsati/ekavīsā*

30. *tiṁsati/tiṁsā*

40. *cattāḷīsati/cattāḷīsā*

50. *paññāsati/paññā*

60. *saṭṭhi*

70. *sattati*

80. *asī*

90. *navuti*

100. *satam*

1000. *sahassam*

Appendix—B

The number-words after *daśa* and before *śatam*, as we have said before, are all compound words. One of the peculiarities of the style of the Vedic language is that it explains some of the compounds. Take, for example, the compound *acyuta-cyut*, made up of two components viz. *acyuta* and *cyut*, which latter is a root-noun from the root -√*cyu* 'to move'. Though theoretically the compound can be dissolved with the help of all the seven *kāraka*-relations of the *prathamā vibhakti*, the *dvitīyā*, *tṛtīyā*, *caturthī*, *pañcamī*, *ṣaṣṭhī and saptamī* , the Veda does not intend all these relations; it intends only the relation of the *dvitīyā-vibhakti* or the acc. case in dissolving the compound; *cf.* for the dissolution of the compound with *dvitīyā*-case, RV. 1.85.4, 167.8, 2.12.4, 24.2, 3.30.4, 6.31.2 etc. The number-words also being compounds are dissolved in many places in the texts of the *saṁhitās*. The present Appendix B aims of collecting all such dissolutions and present them to have the view at a glance. In all 69 number-words between *dośa* and *śata* are stated in the nine Vedic saṁhitās, out of which only 12 are dissolved in different places.

A study of the dissolution of the Vedic compounds forms part of a bigger project of mine entitled 'A concordance of Vedic Compounds Interpreted by Veda' and actually a volume under the same title is already published by the CASS, University of Poona, Pune-412007 in 1989 itself.

It should be noted that though these compounds are words, they convey a mathematical meaning and not the general linguistic meaning signifying some concrete, physical thing. Since the very purpose and nature of these word-structures are different from the other non-mathematical word-structures, the order of the two constituent members of the mathematical compounds is to be reversed while interpreting and transforming them into symbols. What is *pūrva-pada* in the compounds linguistically is to be taken as *uttara-pada* mathematically; and what is *uttara-pada* linguistically is to be moved to *pūrvapada* mathematically. Thus, in *pañca-daśa, pañca is* the *pūrva-pada* and *daśa* the *uttarapada* from linguistic point of view; but mathematically, the *pūrva-pada pañca* becomes the *uttara-pada* and the *uttara-pada daśa* moves to the *pūrva-pada.* Though linguistically the order of the constituent members is *pañca* and *daśa,* mathematically the order changes to *daśa* and *pañca;* in symbols, *pañca-daśa* (=51) = *daśa-pañca* = 15, the symbol 1 for *daśa* indicating the first series of the two-digit numbers. The basis on which such an interpretation of the mathematical language is to be done is the axiom *aṅkānām vāmato gatiḥ*, which is followed throughout the world right from the ancient times of the Vedic civilisation.

It will be seen from the following dissolutions of the compound number-words that they are not dissolved in the same way and on the same pattern, just as the non-number-word-compounds are not dissolved in the same way and on the same pattern. We have, therefore, to divide the number-word compounds and their dissolutions into two types: (i) those number-words which signify the numbers between *ekādaśa* and *nava-navati* and (ii) those which signify the numbers 200, 300, 4000, 5000 etc.

If we examine the dissolutions of the number-word compounds in the light of the above division, we find that the number-words signifying numbers from 11 to 99 are dissolved as the *dvandva* compounds with the meaning of *ca* (called *itaretara dvandva*), the words signifying numbers above 100, are dissolved as *dvigu tatpuruṣa* or *dvigu samāhāra* as it is called; *cf.* the Pāṇinian *sūtras,*

2.2.29 (*cārthe dvandvaḥ* for *dvanda*) and 2.1.52 (*saṁkhyāpūrvo dviguḥ*, for *dvigu*). Thus, the compound number-words signifying numbers between 11 and 99, viz. *ekādaśa, ekaviṁśati,* etc. are dissolved as *dvandva* compounds as *ckā ca daśa ca, ekā ca riṁśatiś ca etc.* As different from this type of dissolution, the number-words *tri-śata, pañca-śata* etc. are dissolved as *tri + śata, pañca śata* etc; or to put it traditionally, *trayāṇām śatānām samāhāraḥ triśatam, pañcānām śatānām samāhāraḥ pañcaśatam etc.*

It should be noted that the compound word-numbers are not dissolved in any way other than the above two ways.

Compounds of number-words dissolved in the Vedic texts.

1. एकादश (=11) dissolved as 1 + 10

   एका च मे दश च मे अपवक्तार ओषधे। AV.5.15.1.

   एकेया च दशभिश्च। VS.27.33 = AV 7.4.1 = MS.4.6.2.

2. एकविंशति (=21) dissolved as 1 + 20:

   एकं च यो विंशतिं च। RV.7.18.11.

3. द्राविंशति (=22) dissolved into 2 + 20:

   द्राभ्यामिष्टये विंशती च। VS.27.33 = AV.7.41 =

   द्वे च में विंशतिश्च। MS. 4.6.2.

   द्वौ च ते विंशतिश्च। AV. 19.47.5.

4. त्रयस्त्रिंशत् (=33) dissolved as 3 + 30:

   त्रिंशतं त्रींश्च देवान्। RV.3.6.9 = AV. 20.13.4

   ये त्रिंशति त्रयस्परो देवासः। RV.8.28.1

   ये स्थ त्रयश्च त्रिंशच्च। RV.8.30.2 = KS.35.6 = KKS. 47.7.

   त्रिभिर्देवैस्त्रिंशता वज्रबाहुः। VS.20.36 = MS. 3.11.1 = KS. 38.6

   तिसृभिश्च वहसे त्रिंशता च। VS. 27.33 = AV. 7.4.1 = MS. 4.6.2

   त्रयश्च में त्रिंशच्च मे। AV. 5.15.3.

त्रयस्त्रिंशच्च वाजिनि। AV. 19.47.4

त्रिंशत् त्रयश्च गणिनो रुजन्तः। TS. 1.4.11.1

त्रयश्च त्रिंशच्च। KS.35.6 = KKS.47.7

5. चतुश्चत्वारिंशत(=44) dissolved as 4 + 40

त्वतस्त्रश्च में चत्वारिंशच्च मे। AV.5.15.4.

चत्वारश्च चत्वारिंशच्च। AV.19.47.4

6. पञ्चपञ्चाशत् (=55) dissolved as 5 + 50

पञ्च च मे पञ्चाशच्च मे। AV.5.15.5.

पञ्च च या पञ्चाशच्च। AV.6.25.1.

पञ्चाशच्च पञ्च च AV.19.47.4.

7. अष्टाशीति (=88) dissolved as 8+80.

अष्ट च मे अशीतिश्च मे। AV.5.15.8.

अशीतिः सन्त्यष्टा। AV.19.47.3.

8. त्रिशत (=300) dissolve as 100+100+100 (or even 3 x 100):

अस्य क्रत्वा महिषा त्री शतानि (अपचत्) RV. 5.29.7

त्रिभिः शतैः सचमानावदिष्ट। RV. 5.36.6.

त्रीणि शतान्यर्वताम्। RV. 8.6.47 = AV. 20.127.3.

तत्राहतास्त्रीणि शतानि शङ्कवः। RV. 3.9.9 = VS. 33.7.

त्रीणी शता... असपर्यन्। AV. 10.8.4.

त्री च शता च। KS. 35.6.

त्रिभ्यः शतेभ्यः स्वाहा। TS. 7-2.19.4

9. पञ्चशत (=500) dissolved as 100+100+100+100+100 (or even 5 x 100):

पञ्चभ्यः शतेभ्यः स्वाहा। TS. 7.2.19.6

10. दशशत (=10,00) dissolved as 100 added 10 times (or even 10 x 100):

त्वं ह्यस्त्राणि शता दश प्रति।प RV. 2.1.8.

युक्ता हयस्य हरयः शता दश। RV. 6.47.18.

श्यावीनां शता दश। RV. 8.46.22.

यदा दश शतं कुर्वन्ति। TS. 7.2.1.4

11. चतुः सहस्र (=4,000) dissolved as 1000 added 4 times (or even 4 x 1000)

    गवां चत्वारि ददतः सहस्रा। RV. 5.30.12.

    बभ्रुश्चत्वार्यसनत् सहस्रा। RV. 5.30.14.

12. षट्सहस्र (=6,000) dissolved as 1000 added 6 times (or even 6 x 1000)

    सुषुपुः षट् सहस्रा। RV. 7.18.14.

Notes and References

1. For the knowledge of addition on the part of the Vedic people, see below the section on addition.

2. For the Pāṇinian definition of *prātipadika, cf.* the *sūtra*, 1.2.45: *arthavad adhātur apratyayaḥ prātipadikam*.

3. For the representation of the Sanskrit word-structures, compounded or non-compounded, in terms of the symbols N (for nucleus) and S (for suffix), *cf. M.D. Pandit, CVCIV,* CASS, Pune, 1989, pp. 1-30.

4. For *samāsas, cf.* the Pāṇinian *sūtras, samarthaḥ padavidhiḥ,* 2.1.1. and *saha supā*, 2.1.4; *cf.* also *M.D. Pandit, op. cit.,* pp. 7-20.

5. *cf. K. Menninger, Number-Words and Number-Symbols,* Eng. translation by *Paul Broneer*, MIT, London, 1969.

6. For a comparative study of the etymological principles of Pāṇini, Yāska and *uṇādisūtrakāra, cf. M.D. Pandit,* 'Some Linguistic Principles in Pāṇini's Grammar', IL, 1963.

7. *cf. T.R. Chitāmani, The Uṇādisūtras in Various Recensions,* University of Madras, 1939, p. 132.

8. *cf. K. Menninger, ibid.* p. 75; for other examples *cf.* pp. 86, 111, 113, 114, 170

9. For a detailed discussion on this topic, *cf. M D Pandit, CVCIV,* CASS, 1989, pp. 1-50; *cf.* also *MDPandit,* Pāṇini—A Study in Compound Word-structures, JMSUB, 1963, pp. 71-99.

10: So also for the number-words for 52, 58, 62, 68, 72, 78, 92 and 98; *cf.* B.D. on the *sūtra*, 6.3.49: *evaṁ pañcāśatṣaṣṭisaptati—navatiṣu.*

And we have *dvipañcāśat* and *dvāpañcāśat* (=52), *aṣṭapañcāśat* and aṣṭāpañcāśat *(for 58); dviṣaṣṭi* and *dvāṣaṣṭi* (for 62), *aṣṭaṣaṣṭi* or *aṣṭāṣaṣṭi* (for 68); *dvisaptati* and *dvāsaptati* (for 72); *aṣṭasaptati* and *aṣṭāsaptati* (for 78); *dvinavati* or *dvānavati* (for 92); *aṣṭanavati* and *aṣṭānavati* (for 98).

11. *cf.* BD. *on the* sūtra: *ekādir nañ prakṛtyā syāt.*

12. *cf. K. Brugmanna, ibid.* II. 178; III. 5ff.

13. *cf. K. Brugmann, ibid.* III. 8.

14. *cf. K. Brugmann, ibid.* III. 89

15. *cf. K. Brugmann, ibid.* III. 12.

16. *cf. K. Brugmann, ibid.* III. 14 ff.

17. *cf. K. Brugmann, ibid.* III. 25.

18. *cf. K. Brugmann, ibid.* III. 4.

19. for *pratyaya* as a compulsory category, *cf* Patañjali on Pāṇini, 1.2.45: *pratyayena nitya-saṁbandhāt; nityasambadhāv etāv arthau prakṛtiḥ pratyayaś ca; cf.* also *MD Pandit, Zero in Pāṇini,* CASS, 1990, pp. 10-15 *cf.* also *MD Pandit, CVCIV,* CASS, 1989, pp. 30-35 for gender and number, *cf. MD Pandit,* 'Formal and Non-Formal in Pāṇini,' ABORI, 1975, pp. 49-79.

20. For closing and non-closing morphemes in Sanskrit and Pāṇini's grammar, *cf. MD Pandit, CVCIV,* CASS, 1989, pp. 5-15.

21. *cf. K. Menninger, ibid.* p. 9-11, 18-20.

22. For detailed discussion of this topic *cf. K. Brugmann, ibid.* pp. III. 52 ff.

23. For number—words as adjectives, *cf. Karl Menninger, ibid.* pp. 18-20, also 27-29; 11-12; 48-49 81-82; 453, 454; for a detailed discussion on the declension of the numerals, *cf. A A Macdonnel, A Vedic Grammar for Students,* Oxford University Press, Bombay, 1966, pp. 98-103.

24. *cf. Litāvatī,* verse 14.

25. For details, *cf. MD Pandit, CVCIV,* CASS, 1989, pp. 30-45.

26. *cf. K. Brugmann, ibid.,* p. 4, 49.

27. *cf. K. Menninger, ibid.*, p. 39-42.

28. *cf. A.A. Macdonell, A Vedic Grammar*, 1953, p., 103.

29. For details, *cf. M.D. Pandit, CVCIV*, CASS, 1989, p. 35.

30. This method of factorisation is later on followed and employed, first by the *padapāthakāra* Śākalya in his *padapāṭha* of the RV, and then by Pāṇini, the grammarian, in the technique of *anuvṛtti; cf. M:D. Pandit, Zero in Pāṇini*, CASS, 1990, pp. 4 F. Has Pāṇini borrowed the technique of *ādeśa* from mathematics? *cf. M.D. Pandit, CVCIV*, CASS, pp. 70-80 *cf.* also *J.F. Staal, Euclid and Pāṇini*, PEW.pp. 16-32.

31. This conclusion and procedure is already brought out to light by previous scholars; *cf. A.M. Pandit*, 'Vedic Mathematics' *Nachiketa*, University of Poona, Pune (13th Annual Publication), 1986-87, pp. 44-45; *cf.* also, *Alfred Hooper, Makers of Mathematics*, Random House, New York, pp. 68-69.

32. *cf. Marks Robert W., The New Mathematics Dictionary and Handbook*, Bantam Science and Mathematics, 1967, p. 21.

33. *A.K. Bag, Mathematics in Ancient and Medieval India*, notes only the mention of the odd numbers; *cf.* ibid. Chaukhambha Orientalia, 1979, p. 54.

34. *cf. Marks Robert W., ibid.*, p. 74.

35. *cf. Hans Hademacher* and *Emil Grosswald, Encyclopaedia of Science and Technology*, Vol. 14, p. 698, 1977; *cf.* also *Cohen* and *Ehrlith, Structure of Real Number-Systems;* also *Hardy* and *Wright, Number-Theory*, Cambridge University Press, Cambridge.

36. *cf. Marks Robert W., ibid.*, p. 146 for *śūnya* and p. 156 for zero.

37. *cf. S. Radhakrishnan, History of Indian Philosophy*, Vol. II, p. 335 f.

38. The text used is: *Milindrapañhapāli*, Bauddhabhāratīgranthamāla, 13, Bauddha-bhāratī, Vāraṇasi, 1990; it is translated into Hindi by Svami Dvarikaprasada Shastri.

39. The text is translated and published by *P.L. Vaidya, Madhyamaśāstra*, Buddhistic Sanskrit Text Series No. 10, Mithila Vidyapeetha, Darbhanga, 1960.

40. *cf. A.K. Narain, The Indo-Greeks,* IBH. Vol. I, Introduction, pp. 24-25.

41. *cf. M. Winternitz, History of Indian Literature,* Vol. 2, p. 175.

42. For the Hindi translation, *cf. Svāmi Dvārika-prasada Shastri, ibid.,* pp. 23-25. For English translation, *cf. Rabindra Nath Basu, A Critical Study of Milindapañha,* Firma KLM Privated Limited, Calcutta, 1978, pp. 93-94.

42A. *cf. Madhyama śāstra,* ed. by P.L. Vaidya, Buddhistic Sanskrit Texts, Series 10, Mithilāvidyāpīṭha, Darbhanga, 1960.

42B. *cf.* S. Radhakrishanan, *History of Indian Philosophy,* Vol. I. pp. 662 ff, 1929; *cf.* also by the same author, *History of Philosophy - Eastern and Western,* George Allen and Unwin Ltd. *London,* 1952, pp. 152-218.

43. *cf.* Śivāditya's *saptapadārthī,* ed. by **D. Gurumurti,** Theosophical Publishing House, Adyar, Madras, 1932, p.10; *cf.* also *Tarkasaṁgraha* of Annambhatta, ed. by **Bodas M.R.,** Bombay Sanskrit Series No. LV, 1930, p.6.

43a. *cf.* **D. Gurumūrti,** *ibid.* Introduction, pp. xi - xiii.

44. For detailed treatment of this subject, *cf.* **M.D. Pandit, Zero in Pāṇini,** CASS, Pune, 1990.

45. *cf.* B.D. on the *sūtra,* 1.1.60 = *prasaktasya adarśanam lopasaṁjñaṁ syāt.*

46. *cf. Tattvabodhinī,* a commentary on BD's *Siddhāntakaumudī,* on the *sūtra,* 1.1.60, 1.1.60; *atra dṛśir jñānasāmānyavacanaḥ; darśanam jñānam; tad iha śabdānuśāsanaprastāvāc śabdaviṣayakam sat śravaṇam saṁpadyate ... tanniṣedhaḥ aśravaṇam.*

47. Since the compound is *dvandva,* it requires technically at least two members to form a compound; for details, *cf.* **M.D. Pandit,** CVCIV, pp. 1-5.

48. For details, *cf.* **M.D. Pandit,** *Zero in Pāṇini,* pp.90-117.

49. The source of the verse is *Samayocitapadyaratnamālikā,* a collection of phrases and proverbs to be quoted on suitable or appropriate occasion. The verse is quoted as a joke on the number zero,

which, in spite of having no value of its own, increases the value of the number ten times when written or read with it. The source obviously seems to be secondary and its primary, mathematical source is nowhere tracable in any mathematical work, pre-Vedic, Vedic or post-Vedic. The dictum seems to have been handed down orally and not mentioned specifically anywhere. Yet, it is quoted here because it serves our purpose in the absence of any better and original source from any mathematical work. I am very much thankful to **Dr. R.P. Goswami,** Assistant Librarian, CASS, for finding out the reference for me.

50. For the different systems of writing, *cf.* **A.K. Bag,** *Mathematics in Ancient and Medieval India,* Chaukhambha Oriental Research Studies, Chaukhambhā Orientalia, Varanasi - Delhi, 1979, pp. 52-102 and all the references therein. Which writing system is older, we do not know. Yet, this opposite position between the Vedic-and Persian writings is available in the case of the mythologies of the two also. What are Vedic gods are demons in Avesta and vice-versa, what are Avestan gods are demons in Veda!

51. For a detailed study of the Sanskrit word-structures and their representation in terms of N-S symbols, *cf.* **M.D. Pandit,** C V C I V, pp. 3-34; also, '*Pāṇini* - A study of Non-Compounded Word-Structure; *VIJ,* 1963, pp. 23-38; also 'Pāṇini - A study in Compound Word-Structures', JMSUB (Humanities), 1962, pp. 81-99.

52. One can easily decipher the origin of the Pāṇinian technique of zero or *lopa* in that of the mathematics; and that too, in Vedic times. Pāṇini had definitely borrowed many of his techniques of linguistic description from mathematics; for details, *cf. MD Pandits, Zero in Pāṇini*, pp. 98-108.

53. *cf. K. Menninger, ibid.* pp. 39-49, 113-114, 148, 420; for different systems with the radix 20 and 60, *cf. op. cit.* pp. 57-581 59-701; 377 (for the Mayan vigesimal system) and pp.162-169, and 170 (for the Babylonian sexagesimal system).

54. *cf.* **F. Cajori,** *A History of Mathematics,* The Macmillan Company, New York, p.10

55. *cf.* **M. Menninger,** *ibid.* p.46f.; also **F. Cajori,** *ibid.* pp.63, 65, 68, 114 etc.

56. This idea is found in finger - counting; *cf.* **K. Menninger,** *ibid.* pp. 33-36.

57. For details, *cf.* **Howard Eves,** *An Introduction to the History of Mathematics,* RINEHART and Company, New York, 1953; pp. 9-16; also, **K. Menninger,** *ibid,* p.265 for alphabetic numerals in Hebrew and Greek. The writing of numbers in Devanagarī alphabets like *k, c, ṭ, t, p* etc. is found to have been prevalent in the later Tantric texts in ancient India; *cf.* **B.L. Upadhyaya,** *Prācīna Bhāratīya Gaṇita* (in Hindi), Vajñāna Bhāratī, New Delhi, 1971, 134-135.

58. The six *Vedāṅgas* are ; *śikṣa, kalpa, vyākaraṇa, chandas, nirukta* and *jyotiṣ*. It is with the help of these six auxiliary sciences that Veda is interpreted.

59. Incidentally, the word *vyāsa* from *vi+ās* 'to keep apart' is opposite in meaning of the word *samāsa, 'sam+ās',* which means 'putting together' i.e. compounding.

60. for details of the long journey of zero from India to West, *cf.* **K. Menninger,** *ibid.* pp.400-417, or practically the whole chapter on 'The Westward Migration of the Indian Numerals' (pp.400-421), *cf.* also **A K Bag,** *ibid.* pp.67-76; also *cf.* **O P Jaggi,** *ibid.* p. 132; *cf.* **L.V. Gurjar,** *Ancient Indian Mathematics and Vedha,* 1947; also, **G.B. Makode,** *Prācīna Bhāratīyānchī Gaṇitaśāstrātīl Pragati* (in Marathī), Pune, 1934.

61. *cF.* **D.E. Smitth,** *History of Mathematics, Vol.* I. p.153.

Bibliography

Bag. A.K.; *Mathematics in Ancient and Medieval India*, Chaukhambha Orientalia, Varanasi, 1979.

Basu Rabindra Nath: *A Critical Study of Milindapañha,* Firma KLM Private Ltd., Calcutta, 1968.

Bodas M.R.; *Tarkasaṁgraha of Annambhatta,* Bombay Sanskrit Series No. LV, 1930.

Brugmann Karl; *A Comparative Grammar of Indo-Germanic Languages* (English Translation), Chowkhamba Sanskrit Series LXXXIV, Varanasi, 1972, Vols. I-V.

Burrow T.; *Sanskrit Language*, Faber and Faber, London, 1964.

Cajori F.; *A History of Mathematics* , The Macmillon Company, New York.

Chintamaṇi T.R.; *The Uṇādisūtras in Various Recensions,* University of Madras, 1939.

Cohen and Ehrlith; *Structure of Real Number-Systems.*

D. Gurumurti; *Śivāditya's Saptapadārthī*, Theosophical Publishing House, Adyar, Madras, 1932.

Dikṣita S.B.; *Bhāratīya Jyotiḥśāstra (in M*arathi) Aryabhūṣaṇ Press, Pune, 1931.

Eves Howord; An *Introduction to the History of Mathematics,* RINEHART and Company, New York, 1953.

Firth J.R.; *The Tongues of Men and Speech* , Oxford University Press, London, reprinted 1966.

Gurjar L.V.; *Ancient Indian Mathematics and Vedha* , Pune, 1947.

Hademacher Hans and Grosswald Emil; Encyclopaedia of Science and Technology, Vol. 14, 1977.

Hardy and Wright; *Number-Theory,* Cambridge University Press, Cambridge.

Hockett C.F.; *A Course in Modern Linguistics* , The Macmillon Company, New York, 1958.

Hooper Alfred; *Makers of Mathematics,* Random House, New York.

Jagadguru Shankaracarya Shri Bharati Krishna Tirthaji; *Vedic Mathematics,* Hindu Vishvavidyalay Publication Board, BHU, Varanasi, 1965.

Jaggi O.P.; *Science and Technology in Medieval India,* Atma Ram & Sons, New Delhi, 1977.

Macdonell A.A.; *A Vedic Grammar for Students,* Oxford University Press, Bombay, 1966.

Makode G.B.; *Prācīn Bhāratīyāncī Gaṇita Śāstrātīl Pragati (in Marathi*), Pune, 1934.

Marks Roberts W.; *The New Mathematics Dictionary,* Bantam Science and Mathematics, 1967.

Menninger Karl; *Number-words and Number-Symbols* (English Translation by Paul Broneer), MIT, London, 1969.

Narain A.K.; *The Indo-Greeks*, IBH, Vol. I.

Pandit A.M.; 'Vedic Mathematics', *Nachiket,* (13th Annual Publication), University of Poona, Pune, 1986-87.

Pandit M.D.; *Zero in Pāṇini,* CASS, University of Poona, Pune, 1990.

——; *Līlāvatī of Bhāskarācārya, (English* Translation), Pune, 1992.

——; *A Concordance of Vedic Compounds Interpreted by Veda*, CASS, University of Poona, Pune, 1989.

——; 'Formal and Non-Formal in Pāṇini', ABORI, Pune, 1975.

——; 'Some Linguistic Principles in Pāṇini's Grammar', IL, 1963.

——; 'Paṇini - A Study in Compound Word-Structures', JMSUB, Baroda, Humanities Section, 1962.

——; 'Pāṇini - A Study in Non-Compound Word-Structures' VIJ, Part II, Hoshiarpur, 1963.

Pingree David; Jyoti*ḥśāstra*, A History of Indian Literature, Otto Harasso, Wiesbaden, 1981.

Radhakrishnan S.; *History of Indian Philosophy,* Vols. I & II, George Allen and Unwin Ltd. London, 1930.

——; *History of Philosophy - Eastern and Western* , George Allen and Unwin Ltd., London, 1952.

Russell Bertrand; *Introduction to Mathematical Philosophy* , A Touchstone Book, Simon and Schuster.

Sergiu Al-George; 'The Semiosis of Zero According to Pāṇini', *East and West,* Vol. 17, Nos. 1-2, Rome, 1967.

S.K. Lal; *Female Divinities in Hindu Mythology,* CASS, University of Poona, Pune, 1980

Smith D.E.; *History of Mathematics,* Vols. I & II, Dover Publications Inc., New York, 1951.

Swami Dwarika Prasad Shastri; *Milindapañhapali,* (Hindi Translation), Bauddhabharati, Varanasi, 1990.

Upadhyaya B.L.; *Prācīn Bhāratīya Gaṇit (in H*indi), Vijñāna Bhāratī, New Delhi, 1971.

Vaidya P.L.; Madhy*amaśūstra,* Buddhistic Sanskrit Series 10, Mithila Vidyapeeth, Durbhanga, 1960.

-SANSKRIT TEXTS-

Ṛgveda Saṁhitā (= RV)

Vājasaneyi Saṁhitā (= VS)

Sāmaveda Saṁhitā (= SV)

Atharvaveda Saṁhitā (= AV)

Kāṇva Saṁhitā (= KāS)

Taittirīya Saṁhitā (= TS)

Maitrāyaṇī Saṁhitā (= MS)

Kāṭhaka Saṁhitā (= KS)

Kapiṣṭhla Kaṭha Saṁhitā (= KKS)

Aṣṭādhyāyī of Pāṇini

Mahābhāṣya of Patañjali

Siddhānta Kaumudī

Kāśikā Vṛtti

Tattvabodhinī, commentary on BD's *Siddhānta Kaumudī*.

Praśustapādabhāṣya

Tattvacintāmaṇi

Nyāyamañjarī

Siddhānta Muktāvali

Siddhānta Candrodaya, commentary on *Tarkasaṁgraha*

Tarkasaṁgraha

Madhyamaśāstra

Uṇādisūtras

Nāmaliṅgānuśāsanam

Auṇādikapadārṇava

Śivāditya's Saptapadārthī

Kiraṇāvali of Udayana

Śiva-mahimna-stotram

Vākyapadīya

-Pali texts-

Milindapañhapāli

- Hindi Texts-

Prācīn Bhāratīya Gaṇit

- Marathi texts-

Prācīn Bhāratīya Gaṇita Shāstrātīl Pragati.

Cār Śulbasūtre, by Dr. R.P. Kulka rni, *Prācīn Bhāratīya Jñān-Vijñān, Tilak* Mahārāstra Vidyāpīṭh, Pune, 1974

Līlāvatī by u D. Yeraṇḍolkar, 1889

Līlāvatī by *N.H. Phadke*, Bombay, 1971

Abbreviations

ABORI -	Annals of the Bhandarkar Oriental Research Institute, Pune.
AV -	Atharvaveda Saṁhitā
BD -	Bhaṭṭoji Dīkṣita
BORI -	Bhandarkar Oriental Research Institute, Pune
CASS -	Centre of Advanced Study in Sanskrit, University of Poona, Pune.
CVCIV -	A Concordance *of Vedic Compounds Interpreted by Veda.*
IL -	*Indian Linguistics*, Deccan College, Pune
JMSUB -	Journal of the Maharaja Sayajirao University of Baroda, Baroda.
KāS -	Kāṇva Saṁhitā
KKS -	Kapiṣṭhala Kaṭha Saṁhitā.
KS -	Kāṭhaka Saṁhitā.

MS -	Maitrāyaṇī Saṁhitā.
PEW -	Philosophy - East and West
RV -	Ṛgveda Saṁhitā
SV -	Sāmaveda Saṁhitā
TS -	Taittirīya Saṁhitā
VIJ -	Vishveshvaranand Indological Journal, VVRI, Hoshiarpur, Punjab
VS -	Vājasaneyi Saṁhitā
VVRI -	Vishveshvaranand Vedic Research Institute, Hoshiarpur, Punjab.
VP-	*Vākyapadīya* of Bhartṛhari

Subject-Index